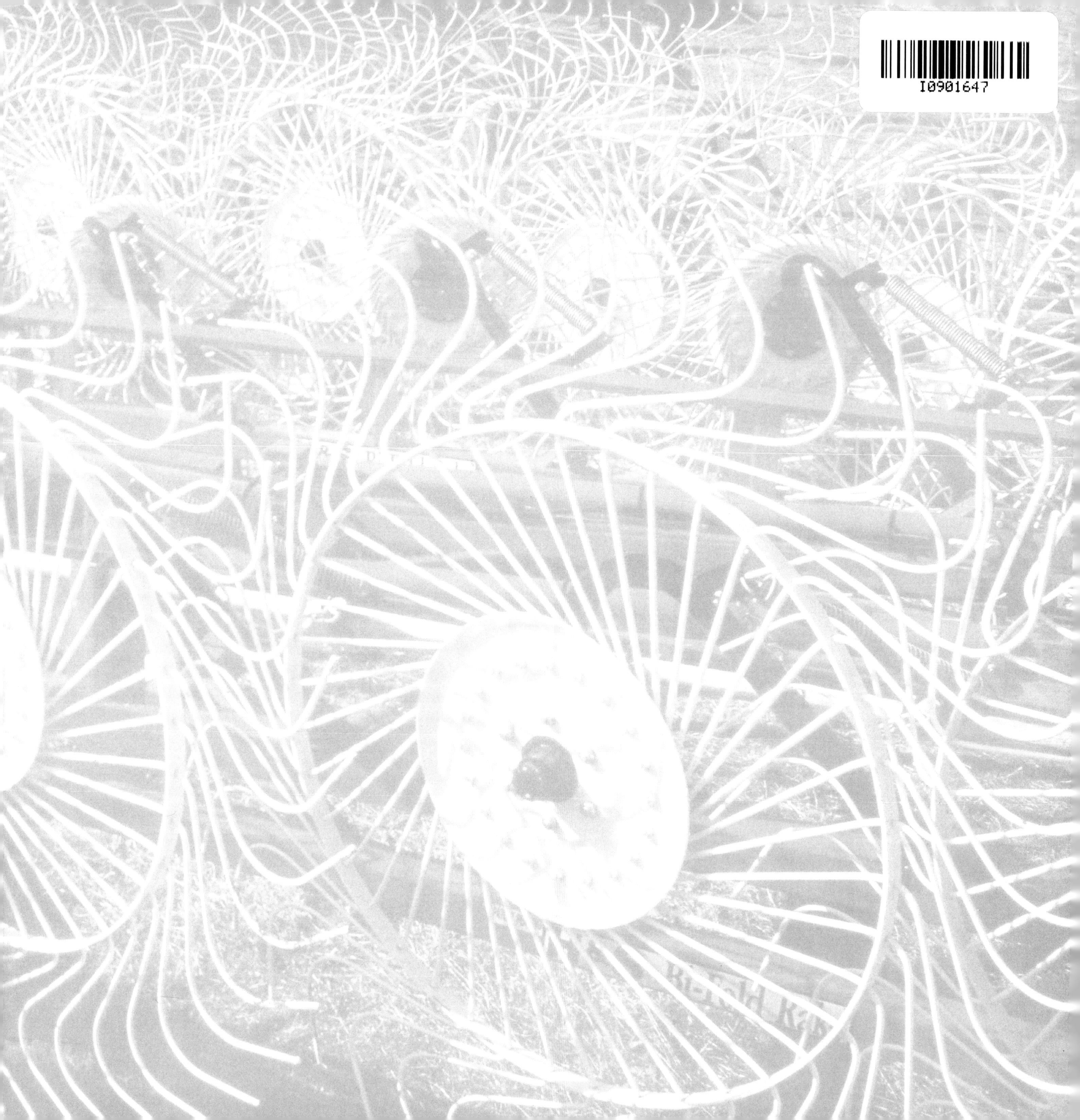

P L A I N S   H I S T O R I E S

John R. Wunder, *Series Editor*

Also in Plains Histories

American Outback: The Oklahoma Panhandle in the Twentieth Century
Richard Lowitt

Railwayman's Son: A Plains Family Memoir
Hugh Hawkins

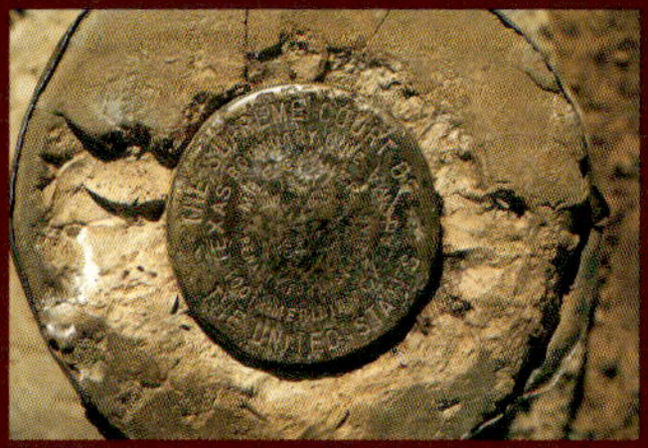

AMERICA'S 100TH MERIDIAN

AMERICA'S 100TH MERIDIAN

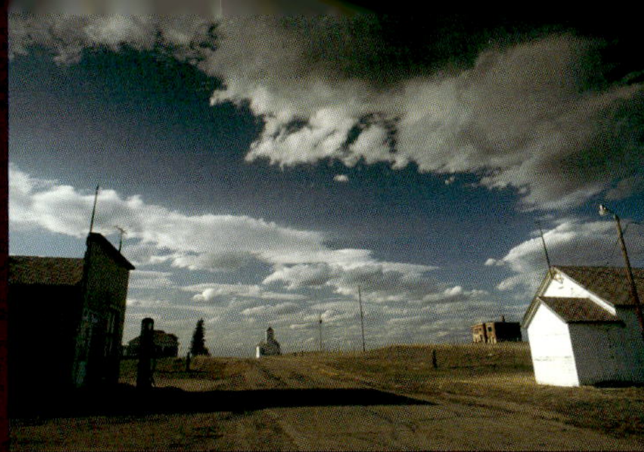

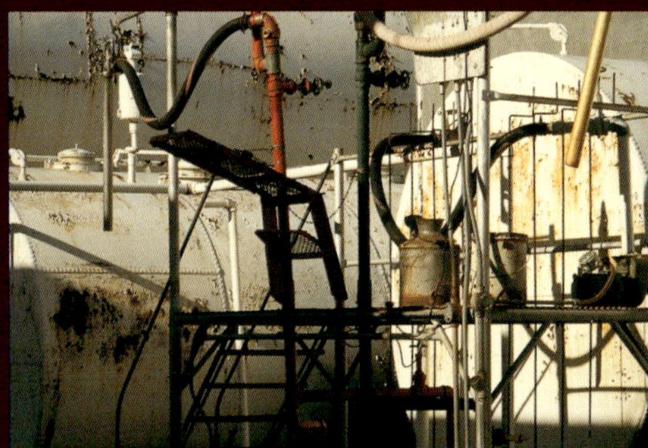

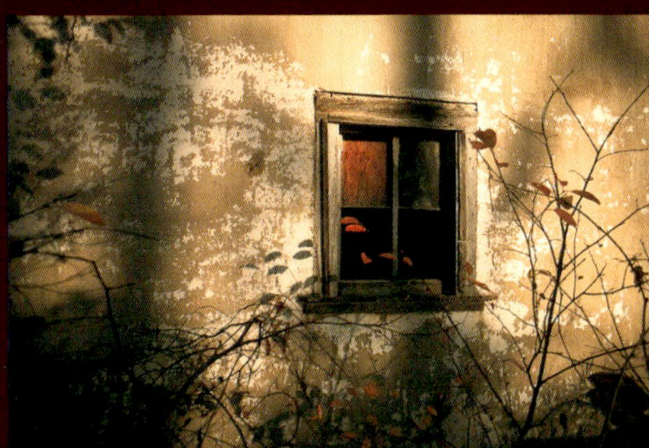

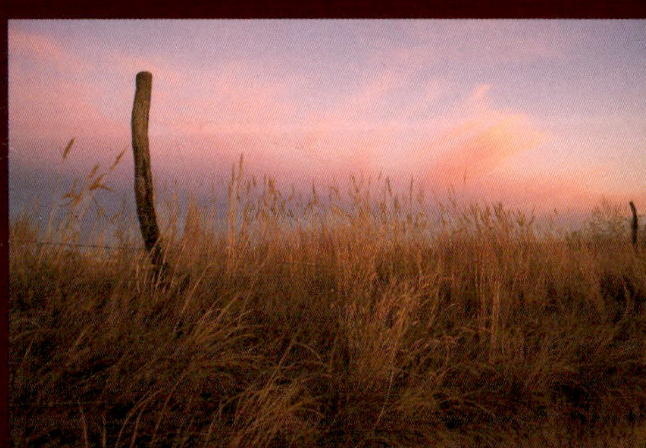

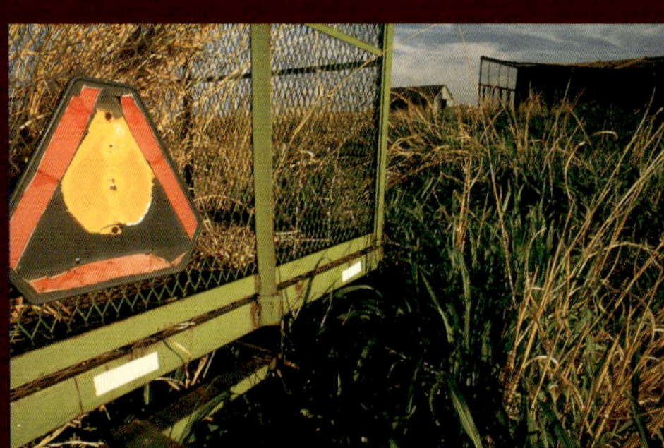

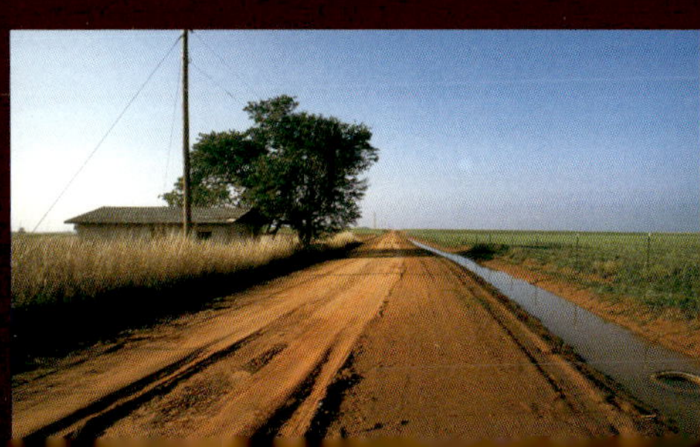

A M E R I C A ' S

1 0 0 T H M E R I D I A N

A Plains Journey

Photographs and text by Monte Hartman

With an Essay by William Kittredge

Foreword by John R. Wunder

Texas Tech University Press

This book is typeset in Scala. The paper used in this book meets the minimum requirements of ANSI/NISO Z39.48–1992 (R1997). ∞

Book design by Barbara Werden

05 06 07 08 09 10 11 12 13 / 9 8 7 6 5 4 3 2 1
Printed in China at Everbest Printing Company
FC

Texas Tech University Press
Box 41037
Lubbock, Texas 79409–1037 USA
800.832.4042
ttup@ttu.edu
www.ttup.ttu.edu

Library of Congress Cataloging-in-Publication Data
Hartman, Monte.
America's 100th meridian : a plains journey / Monte Hartman ; with an essay by William Kittredge.
p. cm. — (Plains histories)
Summary: "Large-format photo study and narratives record journeys along the 100th Meridian from the Canadian to the Mexican borders. Explores life and communities along the vertical line that denotes the nation's geographic center and the shift in annual rainfall to less than twenty inches per year"—Provided by publisher.
Includes index.
ISBN 0-89672-561-8 (hardcover : alk. paper)
1. Great Plains—Pictorial works. 2. Landscape—Great Plains—Pictorial works. 3. Meridians (Geodesy)—United States. 4. Hartman, Monte—Travel—Great Plains. 5. Great Plains—Description and travel. 6. Great Plains—Social life and customs. 7. Natural history—Great Plains. I. Kittredge, William. II. Title. III. Series.
F595.3.H37 2005
917.804—dc22 200501809
ISBN-13 978-0-89672-561-4

TOWN PORTRAIT, Arena, Burleigh County, North Dakota
OILS AND CHEMICALS, Linton, Emmons County, South Dakota
WINDOW, DUNNING, Blaine County, Nebraska
TWILIGHT FENCELINE, Near Nicodemus, Graham County, Kansas
COTTON WAGON, Gould, Harmon County, Oklahoma
COUNTRY ROAD, off Highway 287, Childress County, Texas

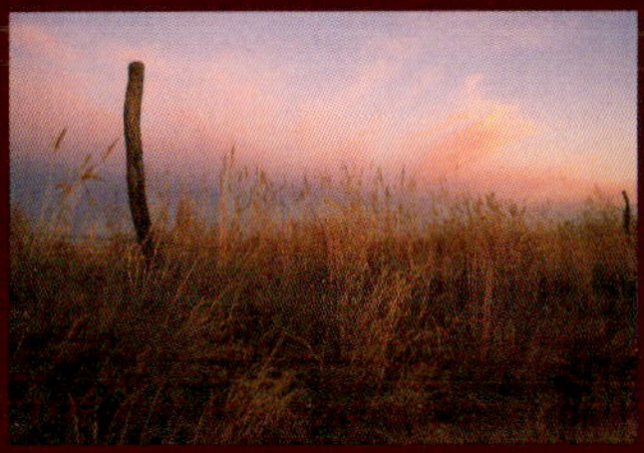

To Lee. Forty years ago she gave me her heart along with the windswept prairie, gifts which proved to be of equal splendor.

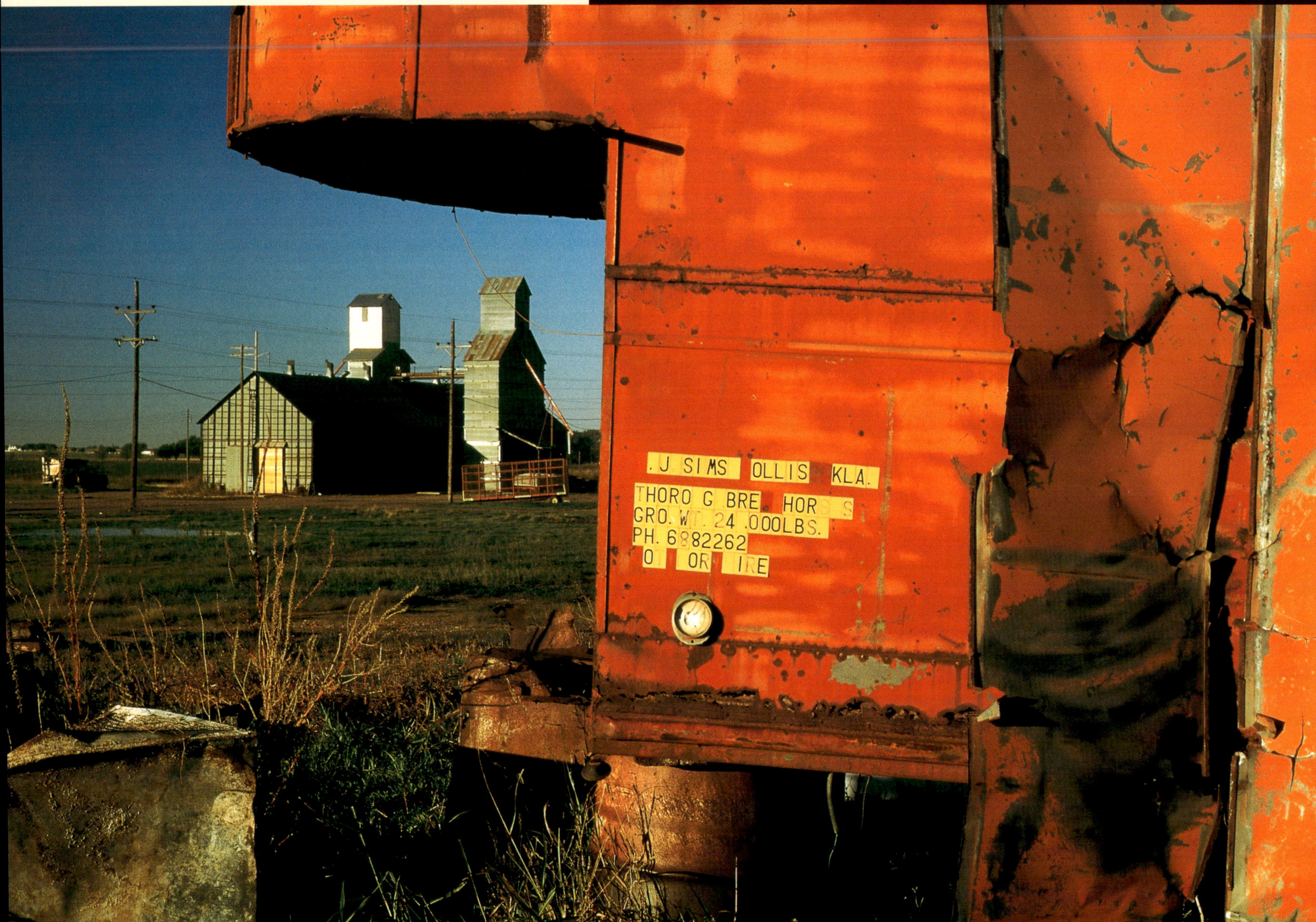

THORO G BRE HOR S

Hollis, Harmon County, Oklahoma

A person is forced inward by the spareness of what is outward and visible in all this land and sky. The beauty of the Plains is like that of an icon; it does not give an inch to sentiment or romance. The flow of the land, with its odd twists and buttes, is like the flow of a Gregorian chant that rises and falls beyond melody, beyond reason or human expectation, but perfectly.

KATHLEEN NORRIS
Dakota: A Spiritual Geography

Contents

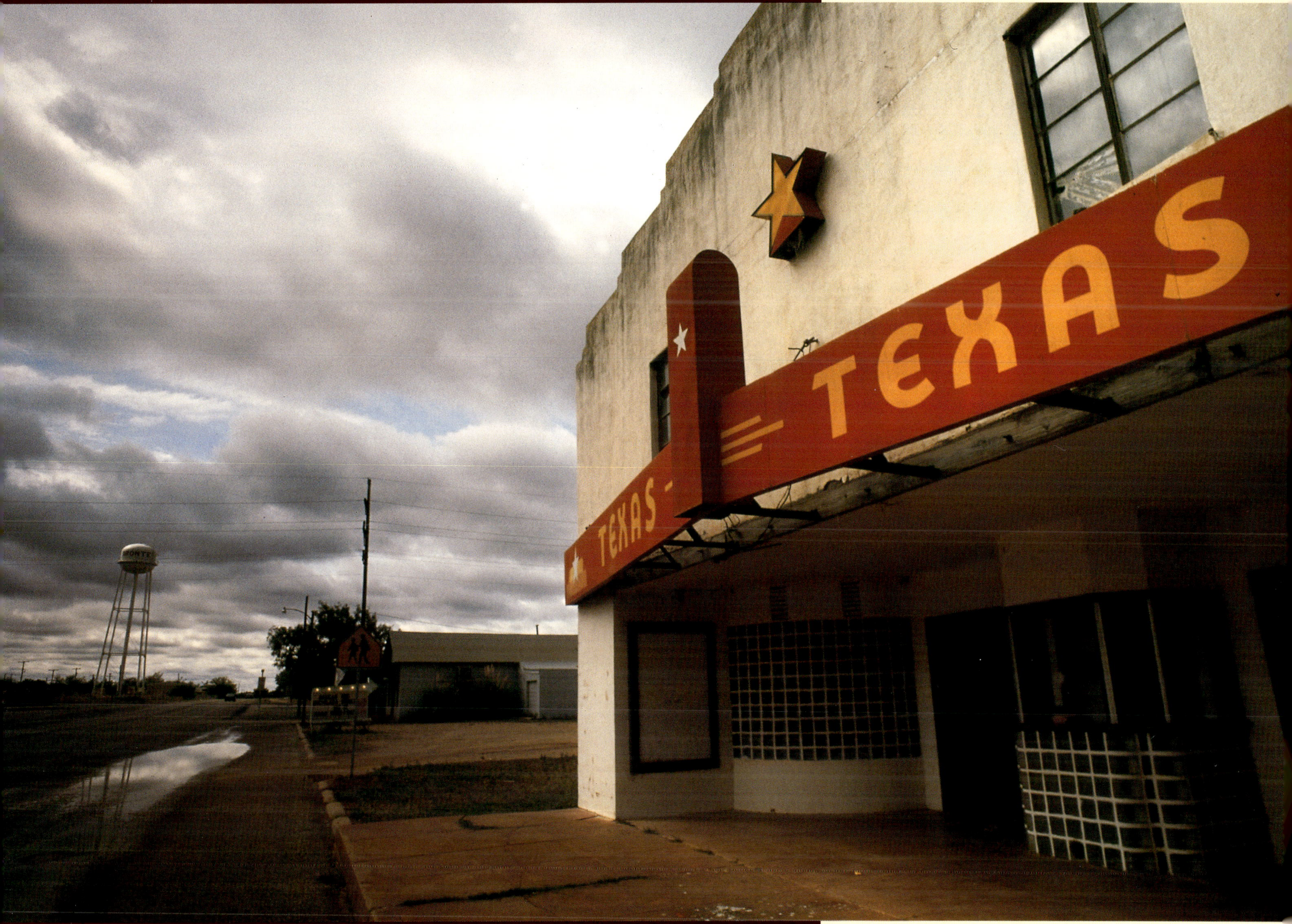

THE LAST PICTURE SHOW
Bronte, Coke County, Texas

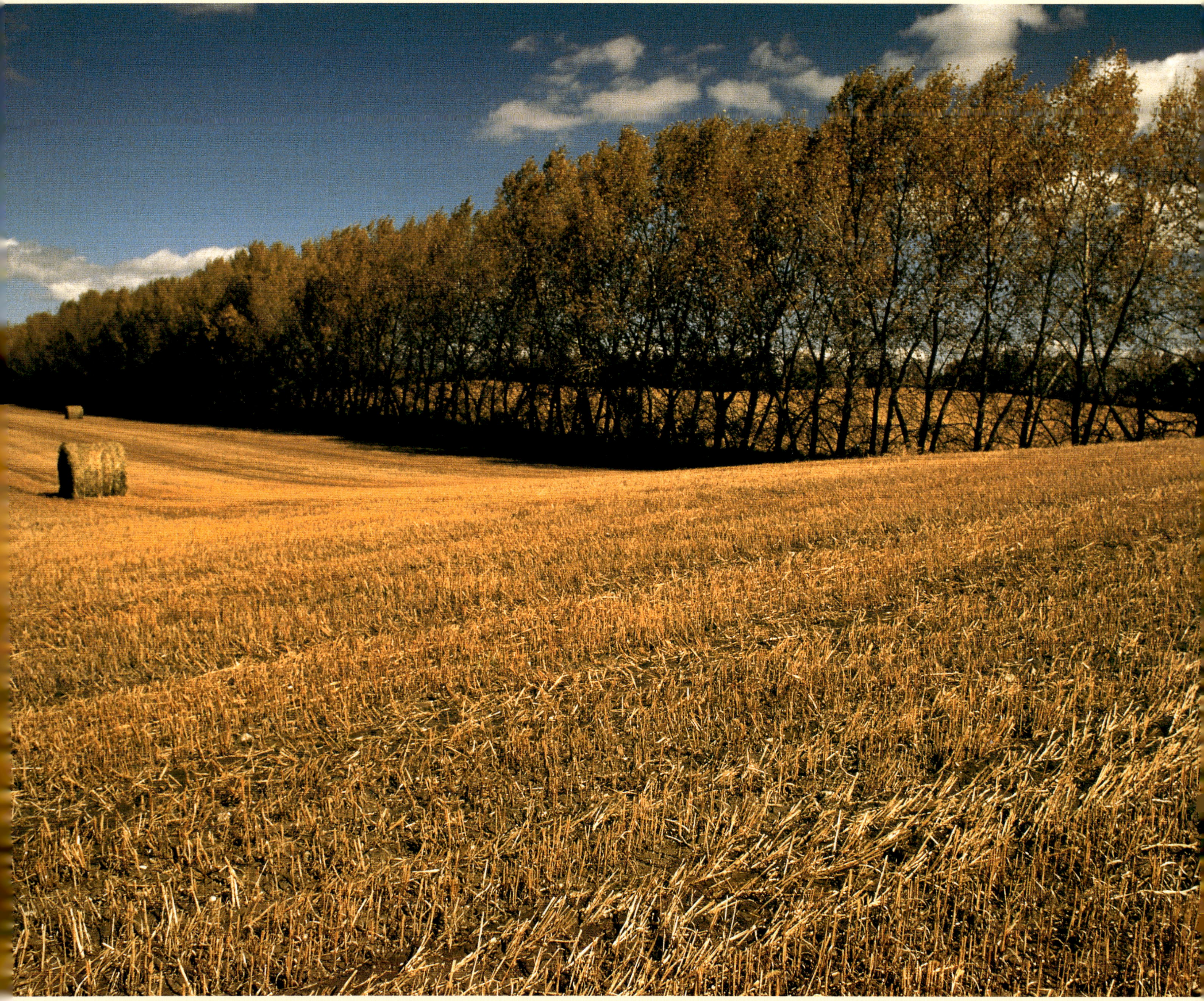

STUBBLE

Turtle Mountain Indian Reservation, North Dakota

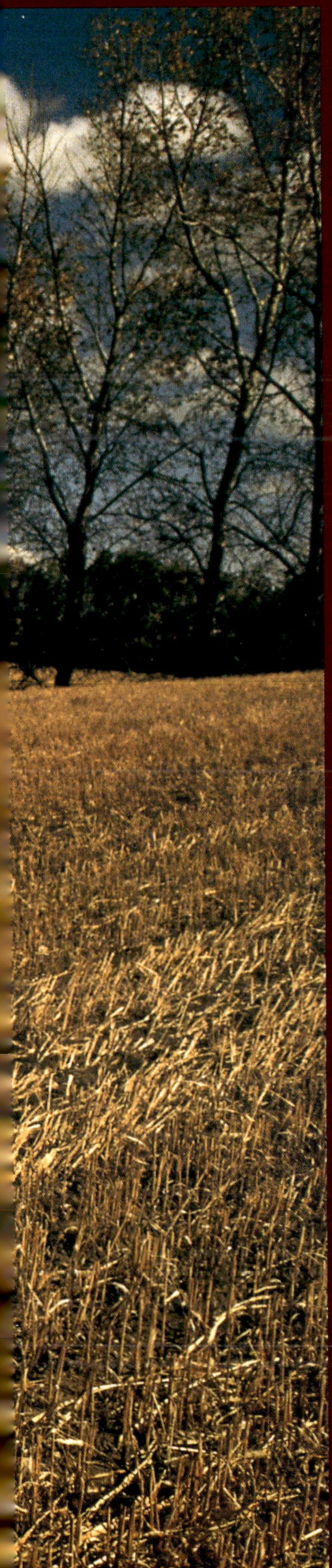

Plainsword

John R. Wunder

Traveling the length of the Great Plains north or south along the 100th meridian means going against the grain. Such a trip defies the predominantly coast-to-coast patterns in transportation that have evolved over the past century. And in many ways, America's east-to-west emphasis, geographically, culturally, and economically, has discouraged interest in the heartland. Wallace Stegner, in his landmark book *Beyond the Hundredth Meridian*, wrote that knowledge of the Plains and especially what lay to the west constituted for those who didn't inhabit the region "a vast Unity of ignorance." He capitalized *unity* to emphasize his point. Stegner sought, as he put it, "the gradual dispelling of the mists, a gradual clarification of the roil of speculation, superstition, guesswork, wishfulness, fear, and misunderstanding."[1] Such is the goal of this book and the new series, Plains Histories, which it initiates.

Monte Hartman conceived of a photographer's journey down the 100th meridian from North Dakota to Texas during a hiking excursion to the northern Rockies. Little did he imagine then that he would eventually embark upon six separate trips, traveling more than 20,000 miles in the vast center of America. As Hartman reflects, he "captured history itself" in his images, and the photos selected for this volume "exploded with meaning." Stark and haunting, geometric and textured, they reveal much about a land and its people. What they are not is time-locked or forlorn; static or drab. Look carefully, and you will sense what Wayne Fields wrote seventeen years ago: "The prairie, in all its expressions, is a massive, subtle place, with a long history of contradiction and misunderstanding." Having perhaps observed the obvious, Fields urged further reflection: "But it is worth the effort at comprehension. It is, after all, at the center of our national identity."[2]

Few people venture the entire length of the Great Plains meticulously, at least not since the Apaches from the north, beginning in the thirteenth century, and the Kiowas from the south, in the sixteenth century. Among those who have recorded their commentary in detail is William Kittredge. His essay, "American Heartline," graces the edges of this volume while it centers on the land and its inhabitants. "The interior of the continent, bisected by the 100th meridian," observes Kittredge, "is a stage where our society has performed, and still performs, giving shape to its purposes." The land from Canada to Mexico is marked, Kittredge reminds us, "as though written upon."[3] History is alive in every visual detail of the Plains.

Others have recorded the Plains in pictures. Recently, Janet M. Neugebauer and Walter McDonald presented *Whatever the Wind Delivers: Celebrating West Texas and the Near Southwest*. Introducing this collection of poems of the Plains and historical photographs, Plainswoman Laura Bush said, "Despite its paradoxical nature, or perhaps because of it, the region commands the respect of its inhabitants. If one thing could be said of those whose lives and livelihoods revolve around the place, it is that they don't simply live on or off the land; they live with it — and thrive."[4] Her words remind us of the paradox that despite the hardships of western experience, some people inhabit the heartland by choice, preferring, if you please, to make their own living, however arduous, in a truly natural environment, surrounded by beauty and free of the crowds, cultural constraints, and pollution characteristic of America's settled areas. These thoughts echo in *The Inhabited Prairie* by photographer Terry Evans with an essay by environmental historian Donald Worster.[5] This pictorial "love song to the prairie," to paraphrase Mari Sandoz's well-known title, and also alluded to by Linda Hasselstrom and many others, begins with a pithy observation by Central City, Nebraska, novelist and photographer Wright Morris. Wrote Morris, of plainsfolk, "There never was a people who tried so hard and left so little behind as we do. There never was a people who traveled so light and carried so much."[6]

N. Scott Momaday, Kiowa Pulitzer prize novelist wrote in *The Way to Rainy Mountain* in 1969 that "Once in his life a man ought to concentrate his mind upon the remembered earth, I believe. He ought to give himself up to a particular landscape in his experience, to look at it from as many angles as he can, to wonder about it, to dwell upon it. He ought to imagine that he touches it with his hands at every season and listens to the sounds that are made upon it."[7] Follow Momaday's behest as you travel the 100th meridian in these pages. Inhabit the landscape, eye it from afar and close up. Wonder and dwell. Listen to the meadowlark and the coyote, the wind and the windmill. And dispel the mists of the Great Plains.

John R. Wunder
Series Editor, Plains Histories
February 2005,
Lincoln, Nebraska

Acknowledgments

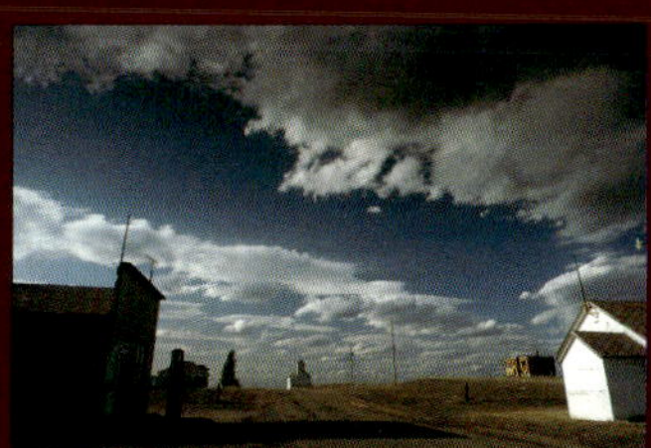

The dream of photographing the knife-edged place where East meets West and publishing the images was realized through the support of many people whose contributions merged into the fabric of this book. Much credit for the richness arising from this effort is theirs; the responsibility for errors and misrepresentations is mine alone.

Acknowledgments start with my wife, Lee, who, in bringing me to the meridian at the ranch pioneered by her family, taught me how the land shapes life for those exposed to the terrible, magnificent, prairie wind. Her sacrifices were manifold, her patience, monumental, her gift, the adventure of a lifetime.

Don Cunningham is singular among those without whom this book would not have been possible. Don's support for the embryonic idea gave me courage, and he was a partner in every step of the process thereafter. His literary talent and knowledge of prairie life made major contributions to whatever quality is manifest in this volume. My debt to him is unbounded.

Enduring thanks to Bill Kittredge for willingness to become spiritual partner of a stranger in this meridian journey. His prodigious talents and sensitivity to the nuances of meridian life enhance this volume hugely, as the poetry of his published writings embellishes our collective being. Fred Woodward's support of the seminal idea shaped the concept for this book. I owe him a great deal.

Enthusiastic thank yous are due Wes Blair for his objective counsel throughout difficult times and Bob Simons, whose friendship and professional advice guided me through the project's contractual and legal shoals. Thank yous also to Harriet Blacker, Alice Acheson, and Susan Eady, as well as to Paul Levin, who helped clear the ground in which the seeds of possibility were nurtured.

Susan Prather, Wes Blair, and Gerry Wasserman assisted in reviewing various texts, Annette Yatovitz' and Don Cunningham's keen eyes shaped the image selection. For their patience and careful scrutiny, I am greatly indebted.

Curatorial assistance during the major exhibits of 100th Meridian Project photography was expertly provided by Martha Kennedy, former Librarian at the Center for Great Plains Studies (University of Nebraska/Lincoln), Roberta Martin of the Kansas State Museum (Topeka), Dr. Judy Diamond and Deborah Meier at the University of Nebraska State Museum (Lincoln). Cy Lehrer's advice on logistics was invaluable, as was the assistance of Bob Wallace, former Director at the Dawson County Historical Society.

The gracious members of the "Friends of the 100th Meridian" who contributed financially to the project were Wes and Shirley Blair, Rich and Arlene DeLeon, Ron and Marie Murphy, Susan Prather, Hugh Ralston, Murray Schmechel, Bob and Kathy Simons, and Gerry and Moie Wasserman. Other contributors include: Lonis and Lois Wendt, Bill and Marie Blunk, Sandra and Joe Goldberg, J. D. Powers, Beryl and Annette Yatovitz, and Harold and Beverly Zane.

Among those who supported me, I particularly remember Mollie Fischer at the Nebraska State Humanities Commission, Don Rowlinson, Chief Ranger at Cottonwood Ranch State Park in Studley, Kansas, and the members of the Dawson County Art Association. For opening their homes and hearts to a tired traveler, I'm ever grateful to Linda Schmechel, Brian and Belinda Briand, as well as Bob and Sue Wallace, who eventually became my dear friends.

A few of the many names have escaped in the mists of memory but not all. Folks offered hot coffee on freezing mornings, invited me into their homes, or cheerfully offered explanations of local legend. Among the Heartlanders I keenly remember are the late Johnny Anders (Shamrock, Texas) and Bob and Gary Gantz of the Bondurant Grain Company (Ness City, Kansas). In Shattuck, Oklahoma, I received a dissertation on prairie economics from Cinda and Waylan Appel. The Director at the 100th Meridian Museum (Erick, Oklahoma) provided details of the ongoing Texas-Oklahoma border dispute. There also was Lyle Perman and his wonderful family from the Rock Hills Ranch (Lowry, South Dakota), Delane Thom (Gettysburg, South Dakota), bookstore owner Duane Gudgel (Valentine, Nebraska), Marcia Hora and Janet Larreau in the Sandhills town of Arnold, Nebraska, Dennis Moore (Eustis, Nebraska), Charles Hagerman and his wife at the now-defunct "100th Meridian Truck Stop" (Sweetwater, Oklahoma). There was Owen Gregory (Merkel, Texas), Epiphanio ("EP") Ortiz (Anson, Texas), J. Robert Neal (Paint Rock, Texas), Jamie Brite, (then) Postmistress in Telegraph, Texas, Heath Hulse (Leakey, Texas), Maria Rivera, former Mayor of Crystal City, Texas, Jaime Pizana (Laredo, Texas), and probably a hundred other warm, friendly people who responded to a photographer from two thousand miles away who cared about their lives.

My grant sponsor was the Dawson County Historical Society, which provided a vehicle for funding and a venue for exhibiting the photographs. Special thanks to its current Director, Barbara Vondras. This volume was made possible in part by a grant from the Theodore G. Baldwin Foundation (Kearney, Nebraska), and my particular appreciation is extended to Mr. Michael W. Baldwin, its treasurer, for hanging in with me in the long wait for fruition.

Of course, I am grateful for the enthusiasm, artistic sensibilities, and professional judgment of Judith Keeling, Kathy Dennis, and Barbara Werden, and others of the staff at Texas Tech University Press, and to Dr. John R. Wunder, general editor of the Plains Histories series, our guru on all matters pertaining to the Great Plains. Together, these talented and dynamic folks gave coherence to the raw materials of my meridian meandering and produced a beautiful book.

Finally, I must thank the citizens of fifty-six counties in the six meridian states who unhesitatingly welcomed this stranger to their midst. Sharing their lives made this book possible, thereby enriching the lives of all of us.

Introduction

The 100th Meridian

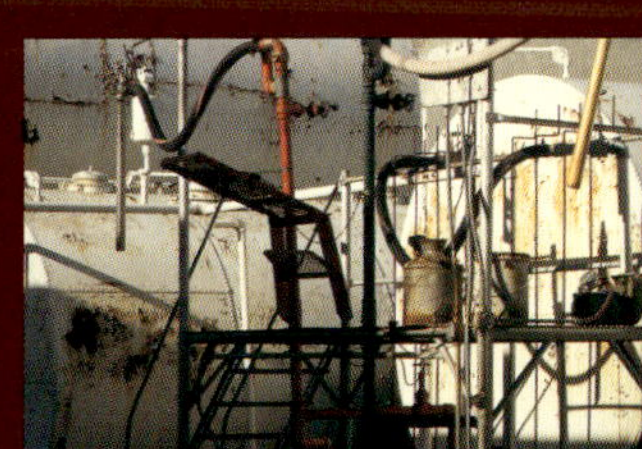

The **100th meridian** in America is a place both ordinary and unique. The invisible 1900-mile north-south line at 100 degrees west longitude marks the separation between, as Walter Prescott Webb and other scholars have described, the "humid" East, with its more than twenty inches of dependable annual rainfall, and the "arid" West, with less than twenty inches of unpredictable precipitation. The place where tallgrass and shortgrass prairies meet and where the flatness of landforms prevails over hills and ridges, this inner boundary of America's Great Plains corresponds with the western extension of the tree line. Marking the definitive limit of eastern farming, the 100th meridian is the imaginary line beyond which the realities of dryness and topography necessitated the introduction of cattle ranching and dryland farming to anchor permanent settlement on the lands stretching westward toward the Rocky Mountains. It is the edge of the West.

In his essay "Living Dry," Wallace Stegner describes the effect of crossing the meridian on travelers following the overland routes — the Oregon Trail, the Mormon Trail, and the California Trail. "Emigrants bound up the Platte Valley on their way to Oregon, California, or Utah, the first targets of westward migration," writes the sage of the North American West, "almost universally noted in their journals that a little beyond Grand Island [Nebraska] their nostrils dried out and their lips cracked, their wagon wheels began to shrink and wobble, and their estimates of distance began to be ludicrously off the mark. They observed that green ceased to be the prevailing color of the

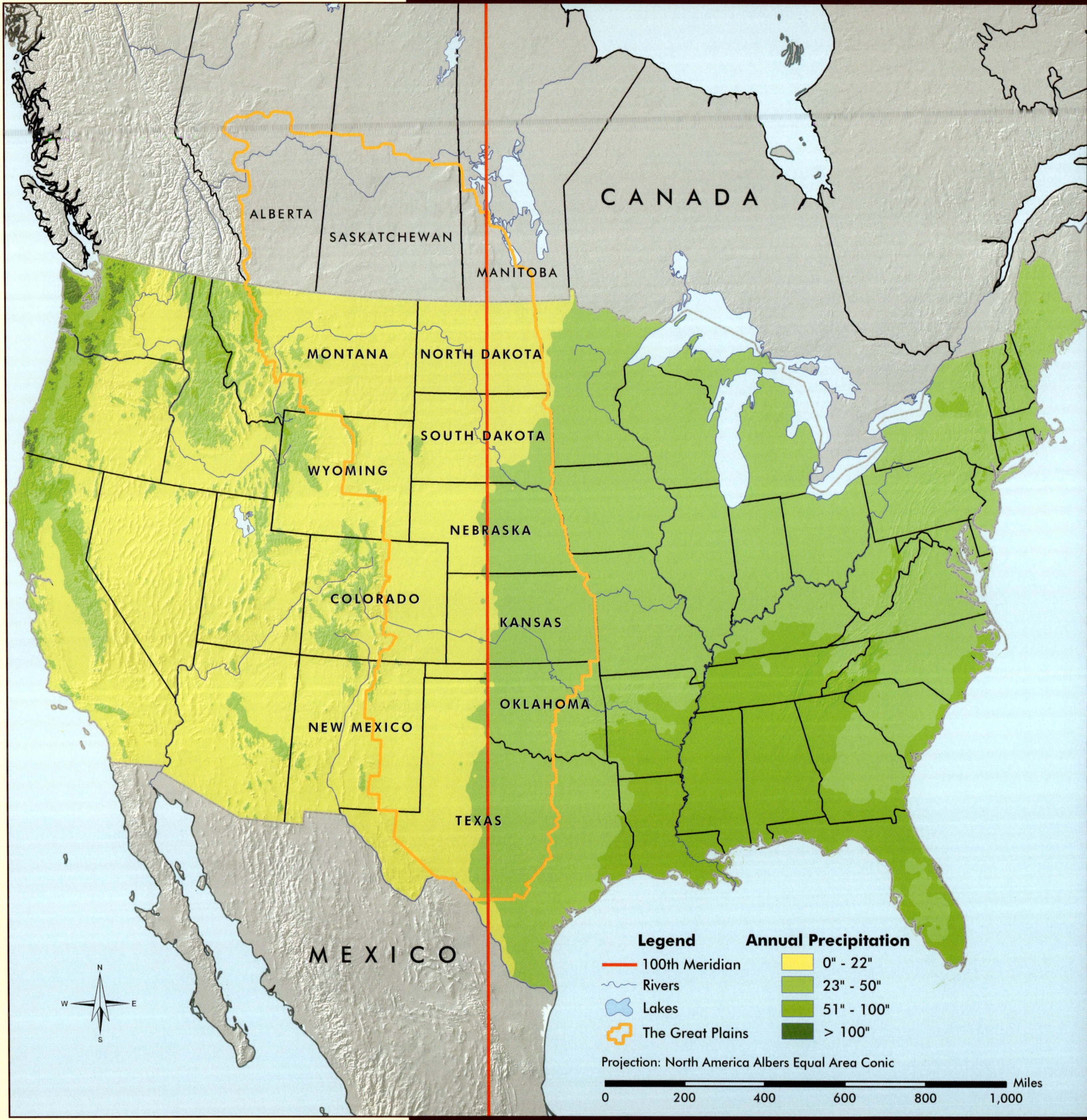

THE GREAT PLAINS AND AMERICA'S 100TH MERIDIAN

Annual precipation and elevation

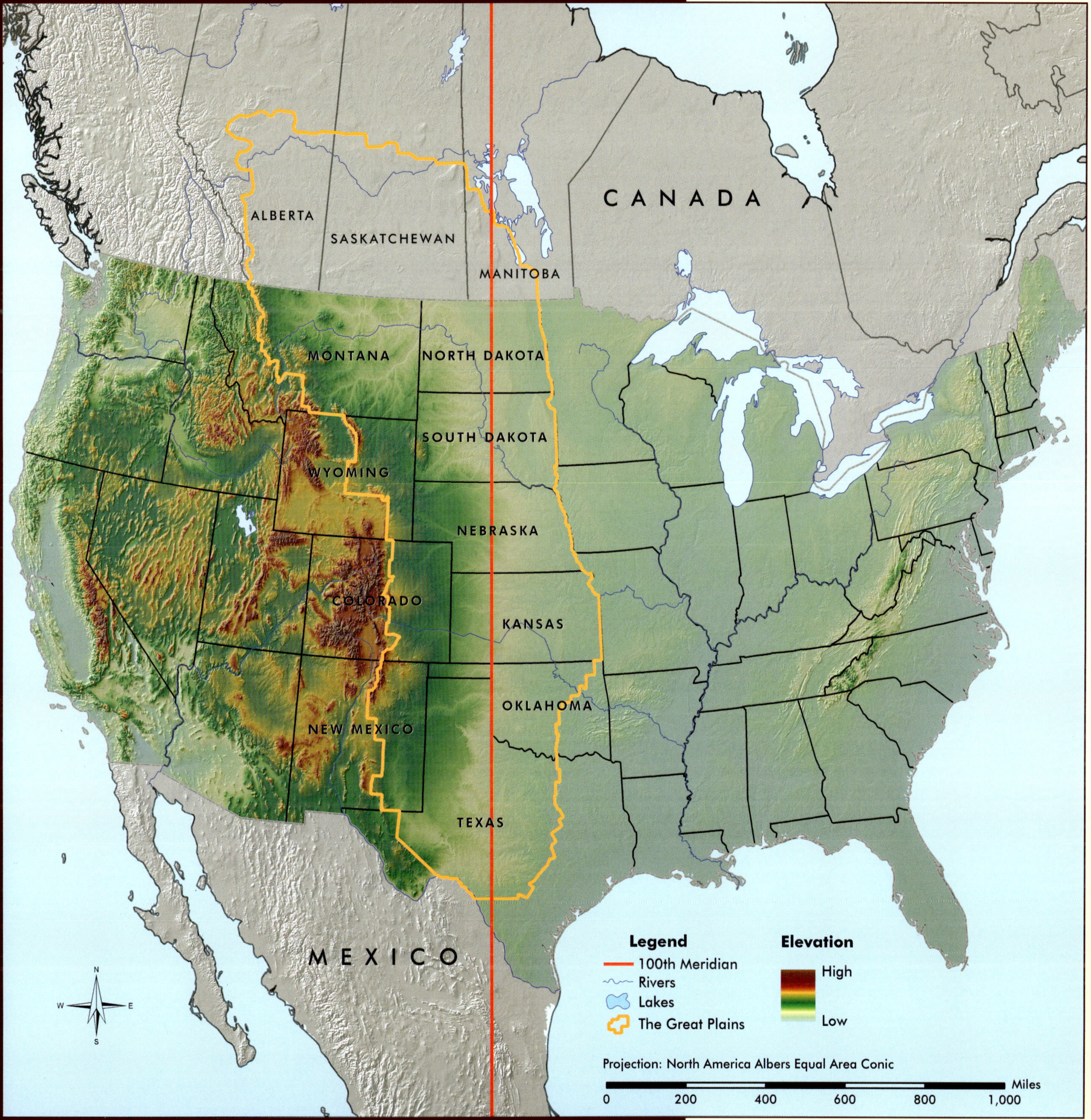

CANADA
ALBERTA
SASKATCHEWAN
MANITOBA
MONTANA
NORTH DAKOTA
SOUTH DAKOTA
WYOMING
NEBRASKA
COLORADO
KANSAS
OKLAHOMA
NEW MEXICO
TEXAS
MEXICO
N
S
E
W
Legend
100th Meridian
Rivers
Lakes
The Great Plains
Elevation
High
Low
Projection: North America Albers Equal Area Conic
Miles
0
200
400
600
800
1,000

earth, and had given way to tans, grays, rusty reds, and toned whites; that salt often encrusted the bottom of dry lakes . . . [and] that the grass no longer made a turf, but grew in isolated clumps with bare earth in between. . . ." Stegner recognized how the environment altered the basic conditions of human presence beginning at or about the 100th meridian.

John Steinbeck also discusses crossing the meridian in his road classic *Travels with Charley*. "Here is where the map should fold. Here is the boundary between east and west," observes Steinbeck about the convergence of the Missouri River with the 100th meridian. "On the Bismarck [North Dakota] side it is eastern landscape, eastern grass, with the look and smell of eastern America. Across the Missouri, on the Mandan side, it is pure west, with brown grass, and water scorings and small outcrops. The two sides of the river might as well be a thousand miles apart."

Maps of railroad development by 1890 show a dense, evenly meshed, omnidirectional web of rail lines that stops at the 100th meridian as though cut by a knife, with only a few slack strands trailing westward. Even these iron symbols of industrialization had to adjust to a new reality when meeting the Great Plains. This map is a paradigm of the differences between East and West — the East characterized by connectedness and proximity, an integration of farm, town, and city in a quilted urbanity. But at the meridian, the closeness and community of the East disappear. Westward, beginning with the 100th, connections are tenuous, spaces are vast, and distance and mobility dominate culture. The meridian is where openness begins.

This Book

Using images of land and townscape to implicitly delineate the struggles with topography and weather that have shaped Plains peoples' lives and a nation's values, this book documents a tough place, a desolate place. This was the last great expanse where masses of people could obtain large chunks of land in the United States. Historical perspective is important to ponder when considering these images. In a region where populations ebb and flow over centuries, these pictures show the most recent depopulation and decay over the last century that, to some, is painful to witness. They also graphically show the kind of losses that can come in the wake of severe forms of "progress." At the meridian is where another agricultural revolution occurred, this time in the twentieth century, when corporate farming, massive equipment applications and investment, and market globalization left the people and land behind in curious ways.

The images documenting the condition of the 100th meridian landscape fall on a continuum between conventional photojournalism and art photography. Their selection is intended to communicate, in artistically and journalistically meaningful ways, a respect for the ethos of *today's* 100th meridian to an urban population largely massed along both American coasts, as well as to Plains inhabitants.

The book begins with a description and a commentary. In "American Heartline," William Kittredge takes us on a personal journey down the 100th meridian during which he reflects on his own past life and seeks meaning from the land and its people. He believes that the key to understanding contemporary attitudes might be found in the solitary and communal traits he has observed in Plains societies, and so he embraces the 100th meridian. Kittredge's companion essay perfectly complements the photographic images.

Over one hundred and twenty color photos taken in the twilight of the twentieth century, from 1993 to 1997, have been selected from more than ten thousand images. Their subjects span the 1,900-mile length of the 100th meridian in the United States, from the Canadian to the Mexican border. Unlike so many visitors traversing America, we shall journey on a north-south route, traveling from the top at the International Peace Garden on the Manitoba-North Dakota boundary to the bottom at the gentle Rio Grande. These thousands of images were accumulated during travels that bisected the states of North Dakota, South Dakota, Nebraska, and Kansas, then crossed western Oklahoma, and continued south through Texas from the eastern margin of the Panhandle to Laredo. They document a six-trip, five-year odyssey totaling over 20,000 miles.

The totality of this book, it is hoped, is a contribution to the cultural history of the Great Plains and middle America. Soak in the meaning from "American Heartline" and then feel it from your observations of the images.

MH

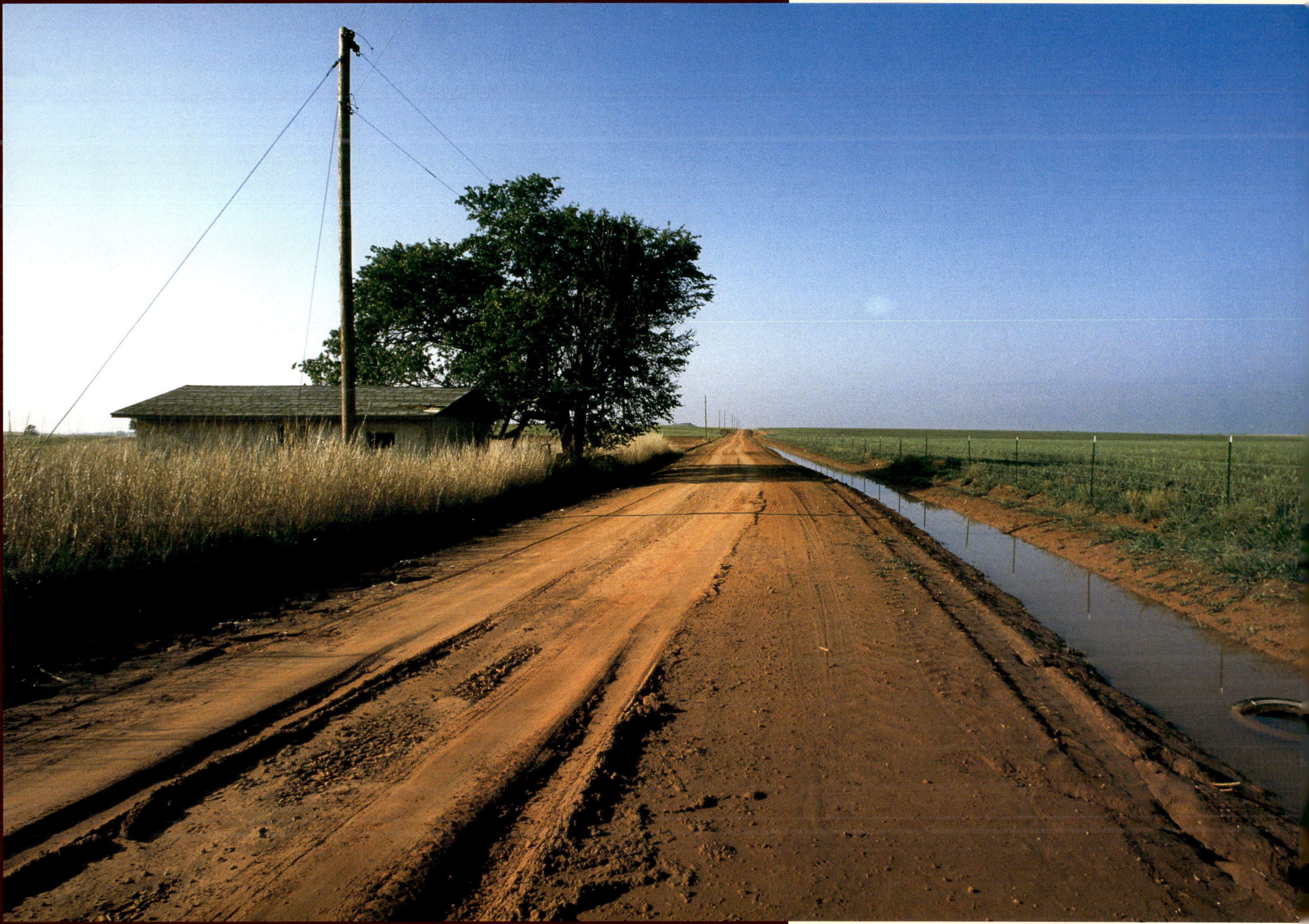

COUNTRY ROAD
Off Highway 287, Childress County, Texas

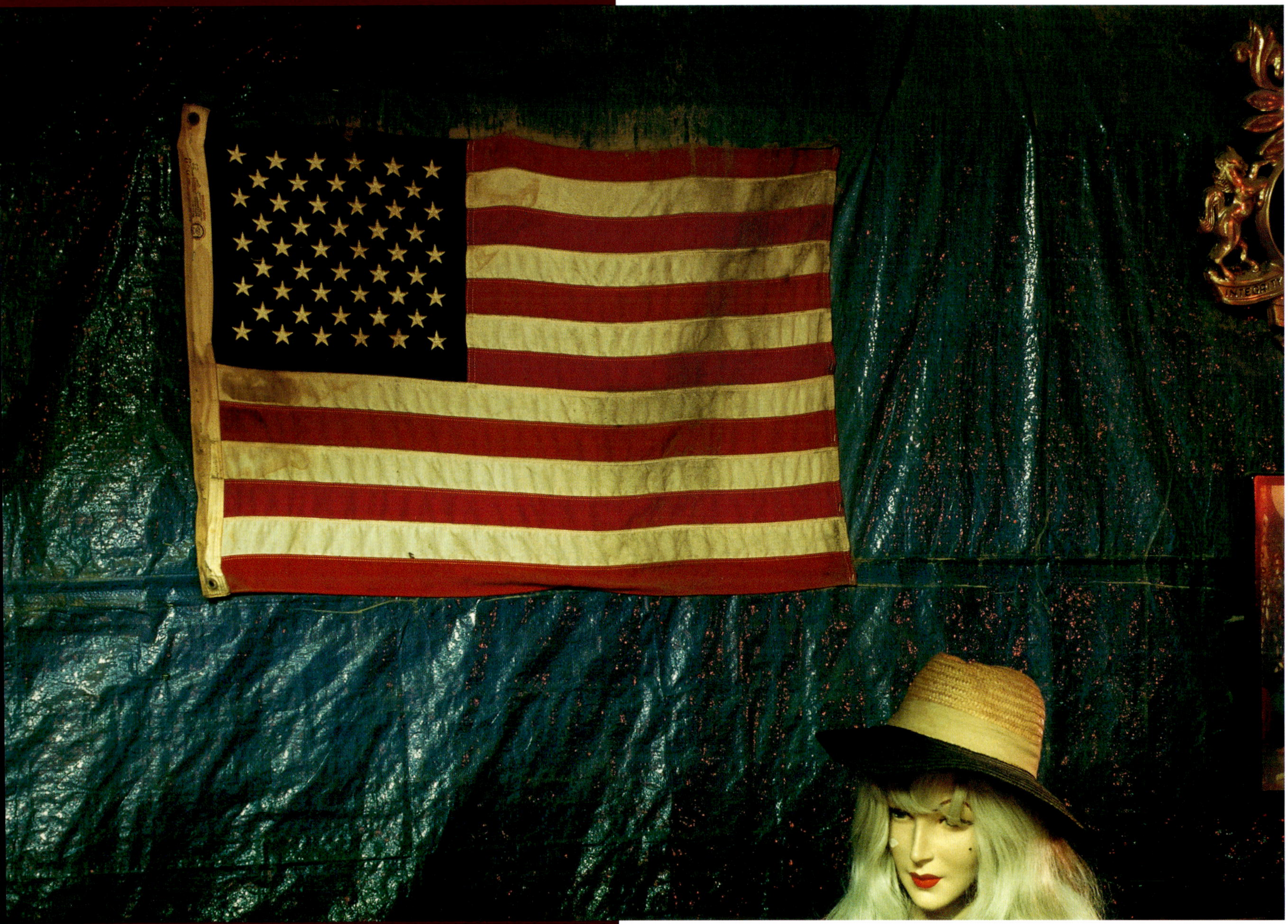

BLOND BOMBSHELL

Murdo, Jones County, South Dakota

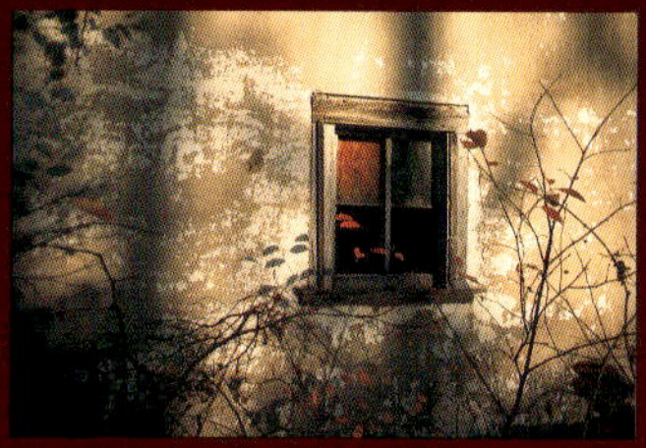

AMERICA'S 100TH MERIDIAN

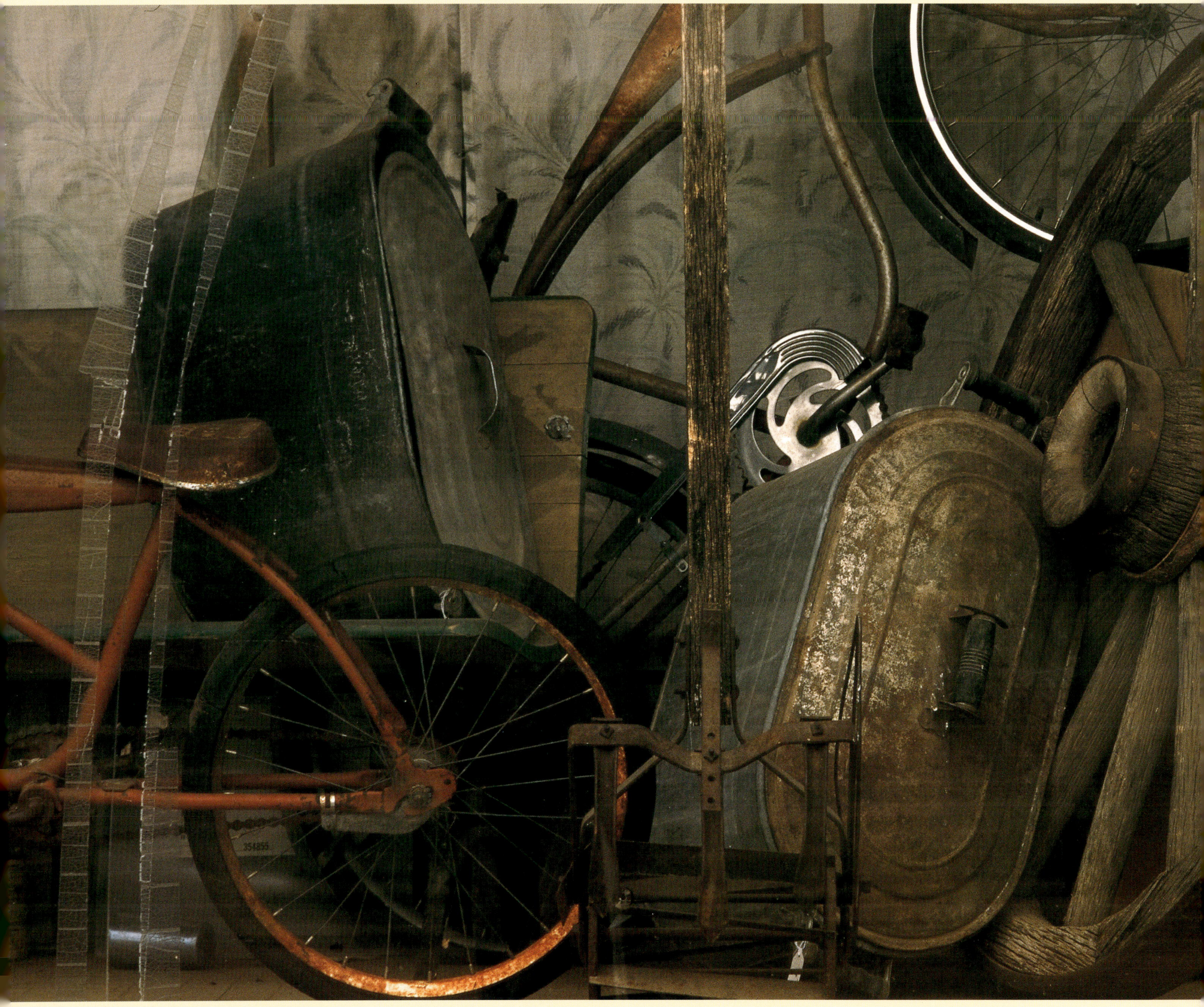

211 CENTER STREET
Holbrook, Furnas County, Nebraska

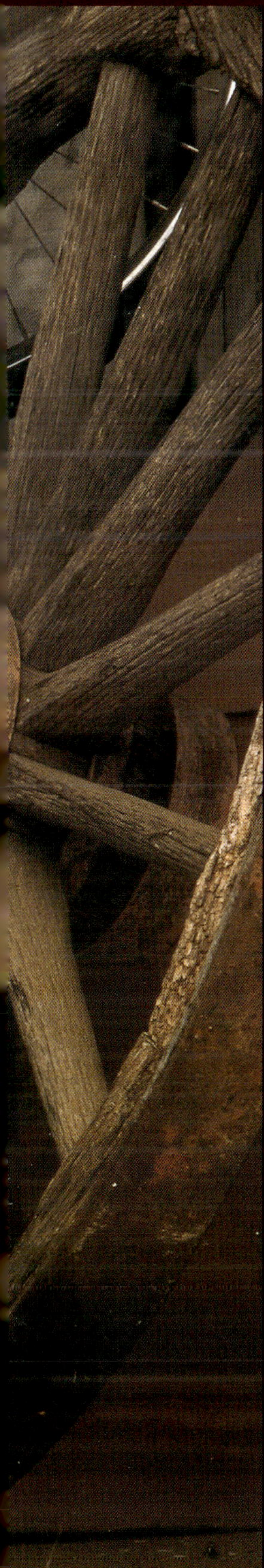

American Heartline

William Kittredge

They hand in hand with wandring
steps and slow,
Through Eden took thir solitarie way.

▪ FINAL LINES OF *Paradise Lost,* JOHN MILTON

It was in the deep expanse of the heartland, according to the thought of generations, that the secret lived. Had to be, or there wasn't any secret.

Our national mirror, the various media, is these days focused almost exclusively on bi-coastal matters. Country life, if it is thought of at all, is mostly regarded as irrelevant to contemporary concerns. Rural people are understood to inhabit an archaic agricultural dreamland. But country life as it has come to exist across the once-upon-a-time grasslands along the 100th meridian is another looking glass, in which we can witness dreams of American homelands as presently enacted.

In June of 1998 I drove the 100th meridian, which runs pretty much down the exact center of the United States, for more than fifteen hundred miles, from Canada to the Rio Grande. I was like an anthropologist, dedicated to spying. I was interested in seeing how the descendants of people who made a stand on what was taken to be an empty stage had used their histories to turn vast territories familiar, into a homeland; I was interested in seeing if their life story of generations, and of place, had become for them interchangeable. And, most centrally, I was interested in seeing if there was any sort of necessary relationship between life in conjunction to vast spaces and societal generosity, if the one kind of openness fed into the other.

My main purpose lay in giving myself a chance to encounter an American culture in which generosity was understood as essential and ordinary. What I imagined was a sweet and open-handed heartland in the center of America. Which turned out, in sadly circumscribed ways, to be quite actual.

Mine was an agenda that began forming decades ago, in the summer of 1968, as I crossed the plains on the run, far from the sagebrush deserts of southeastern Oregon I knew as home. Our ranch was sold, my wife left for California with our children, and I was off to begin a new life among strangers.

Tossing empty beer cans over into the back seat, I wondered if Jack Kerouac and those boys really loved the road. Ray Charles moaned "I Can't Stop Loving You." I was desperate. Give it, I thought, some time. Time heals. Maybe. The vivid sky looked merciless.

A year later, going to a teaching job in Montana, towing a U-Haul and weary from a going away party in Iowa City, I crossed the Missouri in South Dakota and everything changed. Just beyond the bridge I stopped for gas, walked off from my car, and stood with my shadow for company, trying to allow myself to acknowledge that maybe I wasn't always going to be so distant from myself as I'd been during my year in the Midwest. Maybe it was the shattering Western light. I was back in the West, where I thought I knew the rules, where the story about who I was supposed to be made sense, back inside my skin. I vowed that I would never voluntarily find myself back in the Plains again.

My next sight of the Dakotas came on a winter morning, looking into a blizzard from an airliner grounded in Bismarck, watching workmen in hooded goose down coats go stoically about the methodical business of loading luggage and refueling. God, I thought, what drives people to stay in such a place?

Then in the summer of 1982, driving through on my way east, North Dakota turned out to be green and blossoming. The rolling hills and white houses amid green gardens, wetlands mixed with fields of nodding sunflowers, looked to be a vision of the good place on earth, populated by people devoted to taking care. Maybe there, in that heartland, it would be possible to find communities where generosity was the ruling cultural paradigm.

It's my idea that we can't proceed much farther as a world society without rethinking our stories and thus our sense of what is valuable, and that we are presently courting oblivion as we go on destroying gene pools and indigenous cultures. It's my idea that we must turn from a culture mostly driven by an urge to accumulate, to use up if necessary, and invent a society eager to give, to nourish, to take care. It was my thought that I might find enclaves of such a society in the Great Plains.

The World All Before Them

About fifty million years ago, in what is called the Laramide Revolution, mountain chains rose along the western edge of North America. Storms gathered over the Pacific, cooled, and emptied of moisture three times before they reached the interior plains east of the Rockies. In that rain shadow, twenty-five million years ago, a vast, windy north-south run of grasslands came to exist. A wonderland of creatures evolved.

Twelve thousand years ago, around the end of the last Ice Age, the Clovis people appeared, with tools, elegant spear points, and stone hammers. They may have hunted great mammals like the mammoth to extinction. Between eight and nine thousand years ago they were replaced by the Folsom people, who fashioned smaller spear points and hunted the prehistoric bison.

From 4000 B.C. to 2000 B.C. the Plains went through a drought that turned them into a desert. After the rains began again, and the grasses came back, humans resettled in the fertile river valleys, hunting and gathering and grinding seeds, learning to stay put and tend their gardens. This way of living evolved and continued for four thousand years.

Then came the first echoes from Europe, catastrophic diseases, like smallpox and measles, brought by white men. Spanish horses and the white men themselves showed up, beginning with the French traders of the 1740s, followed by Lewis and Clark in 1804. By the middle of the nineteenth century the vast plains had been crossed and recrossed by explorers, and by artists such as George Catlin and Karl Bodmer and John James Audubon, and by gold seekers and farmers, buffalo hunters and the U.S. Army. The great displaced tribes like the Sioux, who for a brief moment were lords on the northern Plains; and the Comanche, the finest horsemen, were horseback warriors who for more than a century amassed an empire from Kansas to Mexico, keeping the settlement of the southern Plains at bay.

This story can be understood as both a bloody hymn to warfare, and a saga about settlement and agricultural people who

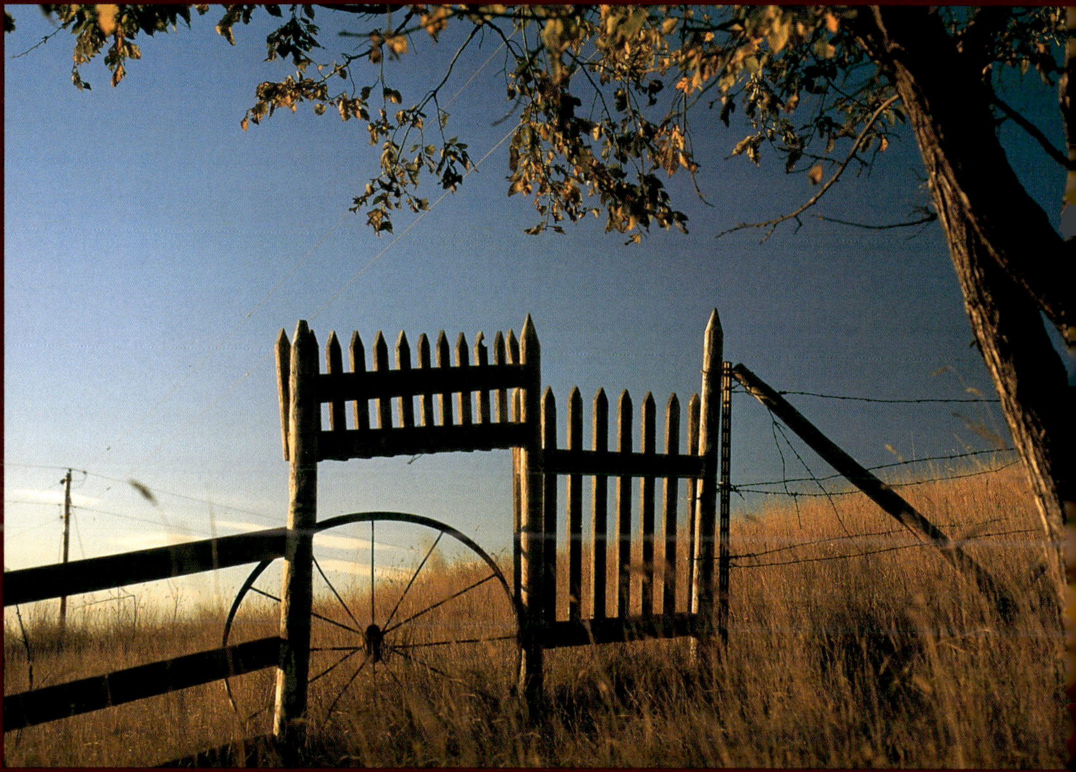

N O R T H   D A K O T A

FENCE
Highway 1804, Emmons County, North Dakota

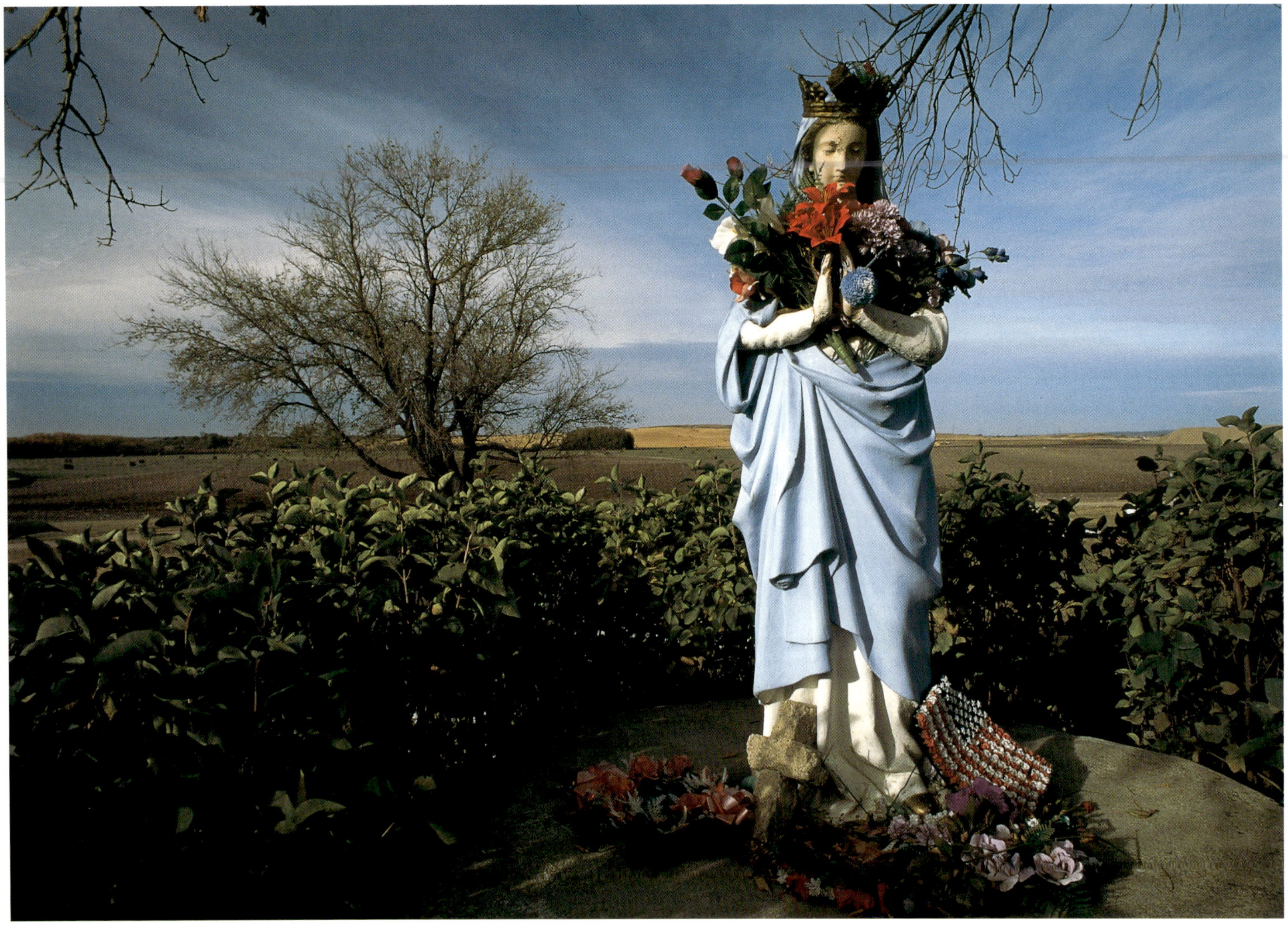

STATUE OF MARY
Turtle Mountain Indian Reservation, North Dakota

came from over the seas, seeking land to farm and freedom. The railroads came, bringing hordes of settlers, families from every county in England and each province of France, schoolteachers from Sweden, and German boys who wanted to be adventurers. The railroads hauled buffalo hides and bones and cattle and wheat away to the east. In every case the native tribes were driven and starved onto reservations.

The nineteenth-century transition from the intimacies of the old European world to prairies where settlers often were utterly alone was typically brutal. What drew the settlers? Promises of an unclaimed garden. What did they find? Howling wilderness. A friend told me that the psychic difficulty inherent in life on the Plains swung on a religious problem: How to live day to day on endless and subtly differentiated landforms. How to study the great skies and the vast towering weather, and the metaphysical infinities they imply, and not give in to feelings of futility and the meaninglessness some people hear ticking in those distances like a bomb. "This land was made for you and me." Or not. In O. E. Rolvaag's novel about life on the plains, *Giants in the Earth,* a young woman thinks "here there was nothing to even hide behind."

The settlement of the Great Plains was part of the enormous diaspora in which 35 million Europeans moved to America between 1815 and 1915. By 1890 nearly a million Scandinavians were settled in Minnesota and the Dakotas, along with thousands of farmers from Russia and Germany. Fields were cleared, ditched, and drained; plow-ground was plowed; houses,

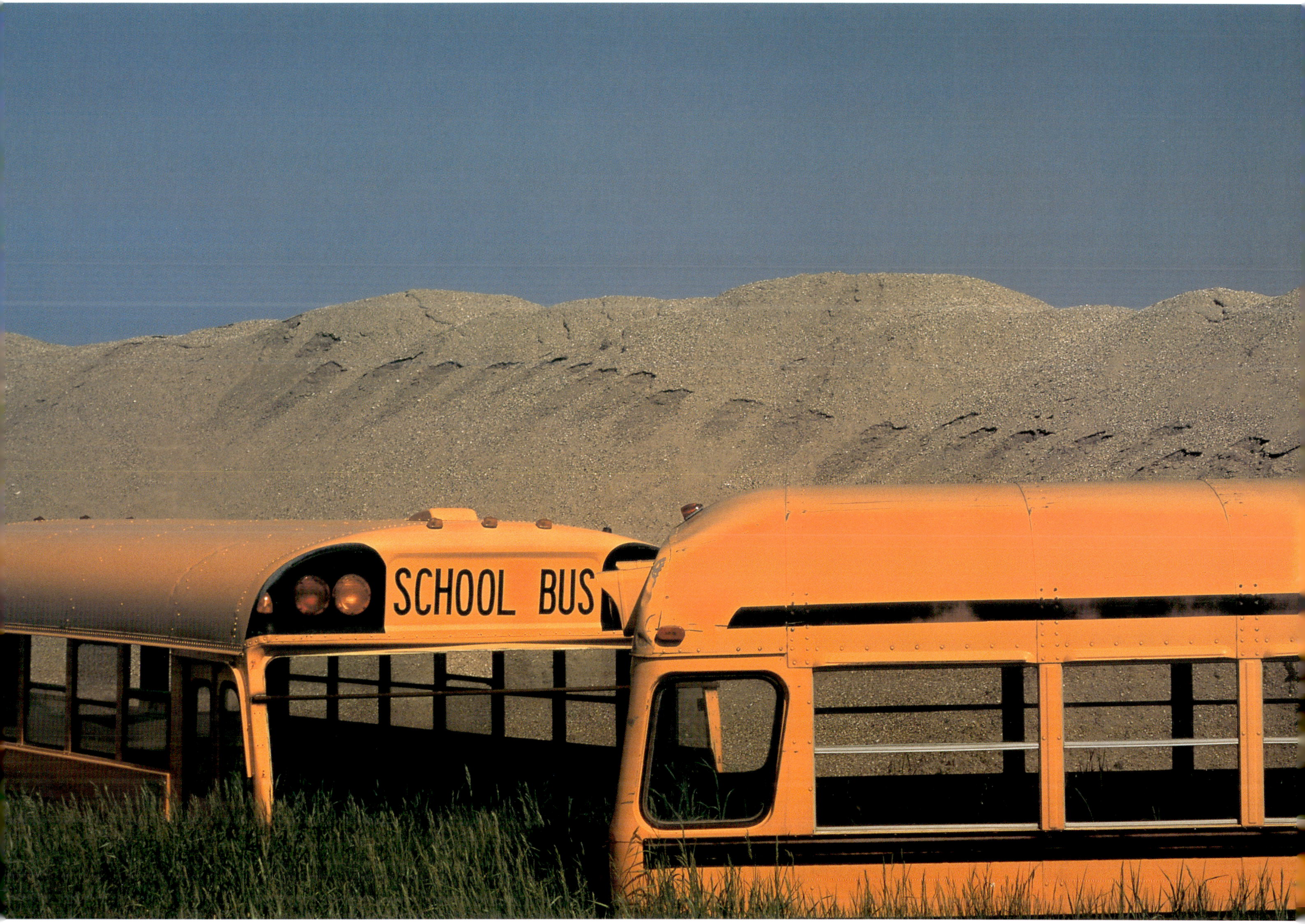

BUS HALVES
Tuttle, Kidder County, North Dakota

VIEW FROM THE TOP
Fessenden, Wells County, North Dakota

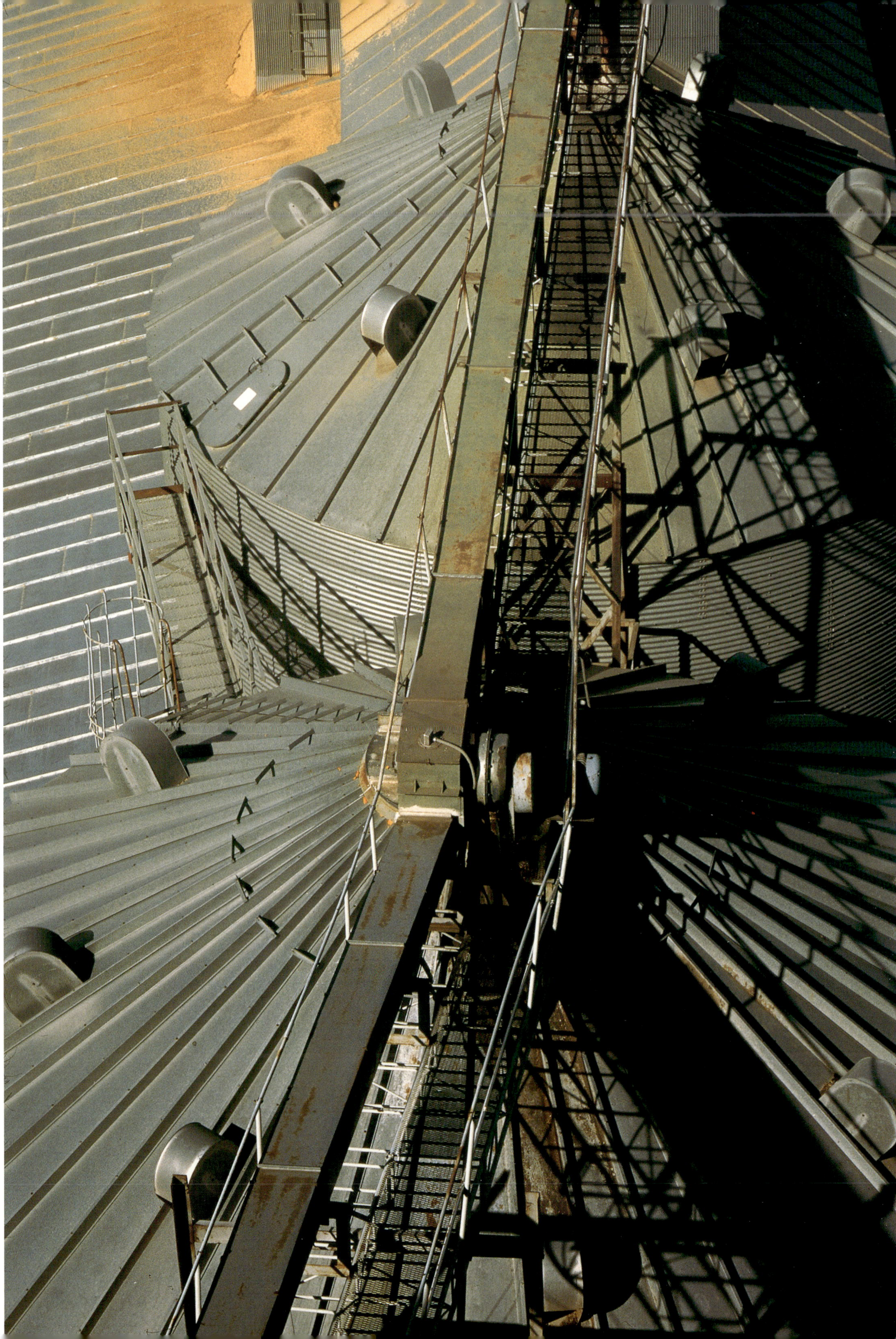

FIRST LIGHT
Fessenden, Wells County, North Dakota

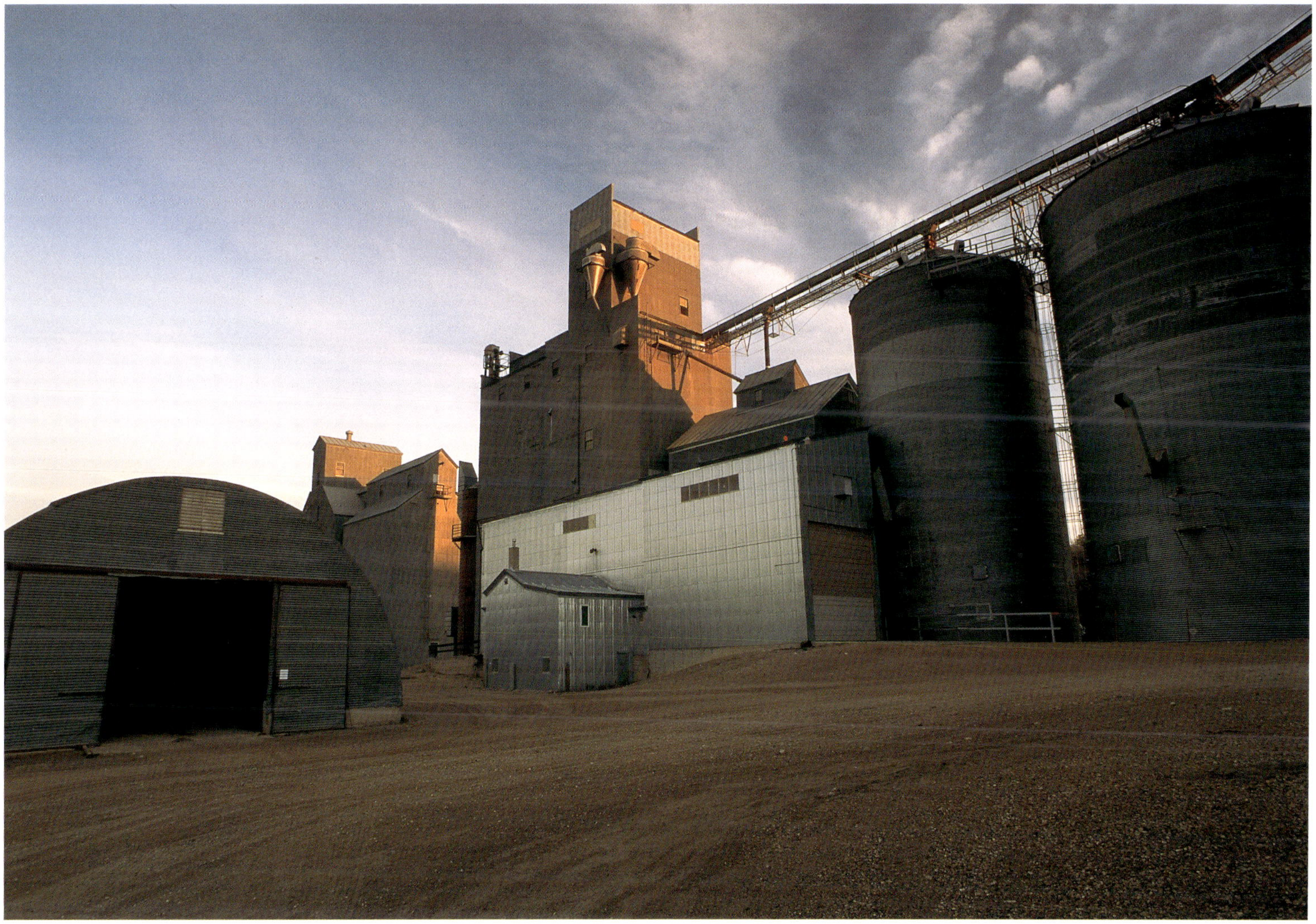

barns, churches, and schools were built; cemeteries were laid out. Generations of people who had suffered to survive were buried. Steel-wheeled steam tractors replaced the work teams. Trees were planted in shelterbelts as protection both from winter winds and psychic isolation. People who longed for community created enclaves where birds and even rodents could nest. Another barrier against infinities.

Some 250,000 settlers came in the years between 1900 and 1915. But, by 1915, during tough economic times, the great tide of settlement turned on the northern Plains, and flowed out, feeding a history of uprooting. Only the toughest remained.

In those difficult years citizens in North Dakota formed a political organization called the Nonpartisan League, and demanded that farmers, not buyers, shippers, and distributors, reap the primary rewards from farming. In 1919 the Nonpartisan League took control of state government in North Dakota, enacting a program of state socialism to control the power of marketing corporations. It was an act of enormous communal will. But, during the 1920s, the value of North Dakota farm property fell by more than a half billion dollars. Five hundred banks failed. Then came drought. Half the population of North Dakota was on relief in 1936. Citizens all up and down the Plains were gripped by antiwar, isolationist sentiments. People in the Dakotas fought to control their lives but couldn't. Their fates, consistently, were determined by outside forces.

But tides eventually turned. Drought ended; World War II brought good grain

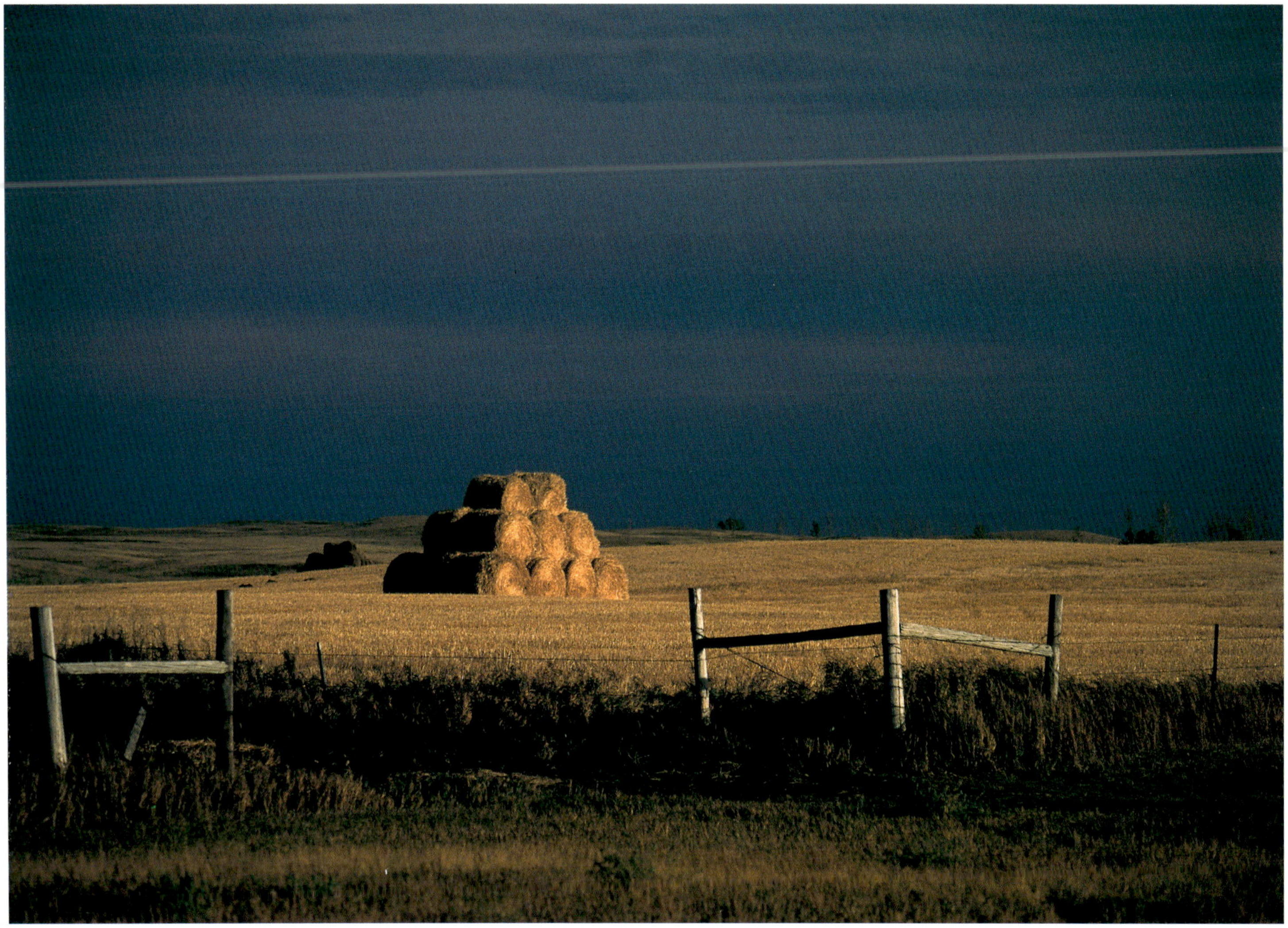

HAYSTACKS AT TWILIGHT
Kidder County, North Dakota

prices; farmers reaped record crops. Electrification, paved roads, automobiles, telephones, and television transformed the nature of rural life. Mechanization drove consolidations of farms (now there are half as many farms in North Dakota as in 1933); small towns died; and the state population dropped by a third between 1940 and 1970. It was a weeding out. A well-off economic class emerged and the concepts central to agribusiness — huge acreages, big equipment, rolling investments, all serviced by multinational corporations — were enshrined as ordinary. The families that had prospered began to isolate themselves into "island communities," caring for their gardens and their kin, and letting the world go to hell on its own.

On the Road Again

Wandering in and out of crossroads towns, the more I traveled, the more I found myself speculating about the price of insistent tranquilities.

Just south of the Canadian border on a brisk day, prosperous farmsteads all around, near the geographic center of North America, I stopped in a cafe with pickup trucks outside for a sweet meal, early corn on the cob and a pork chop. Thick drops of rain were beating the windows.

"Some weather," I said to the farm men who were studying me as I sat down. A couple of them smiled and turned back to local concerns. I honored their conventions, and they resealed the doorway, seemingly complete in their indifference. I propped up a sports page. How had life in what was so widely understood as a promised land led to this walling off?

Soft rain was falling into the Missouri River when I reached Fort Mandan, where

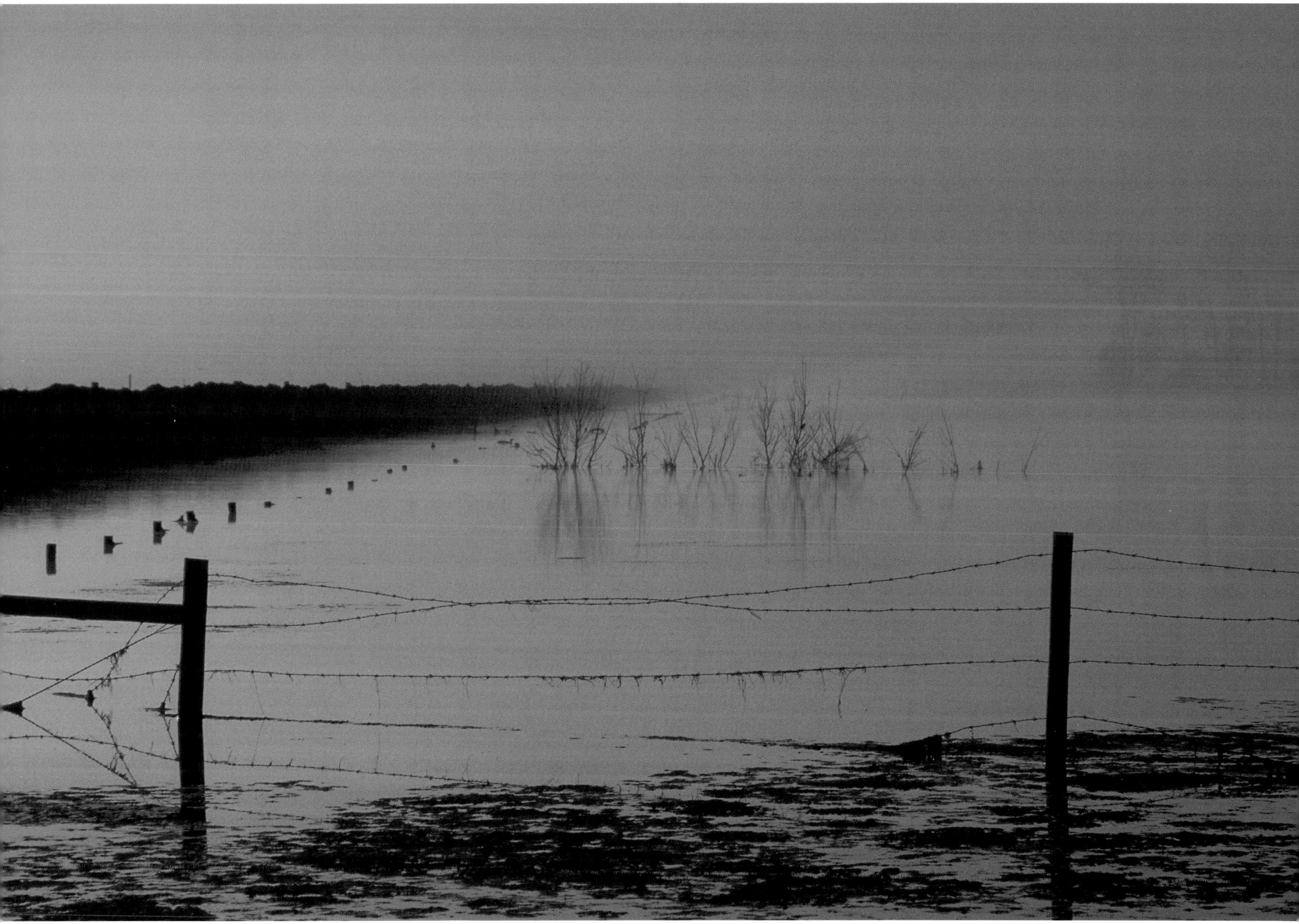

AFTER THE RAINS
Near Napoleon, Logan County, North Dakota

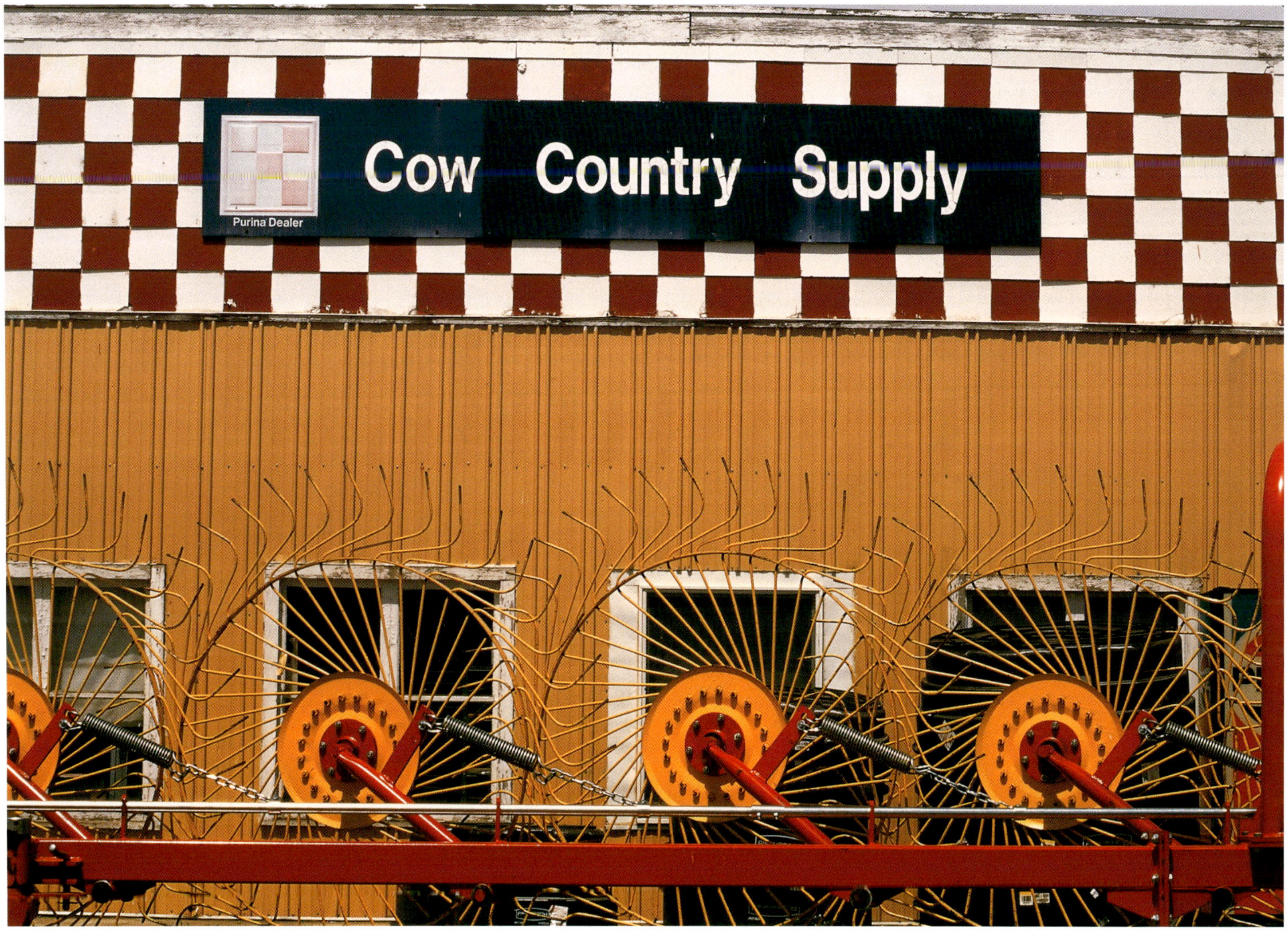

"COW COUNTRY SUPPLY" Rugby, Pierce County, North Dakota

Lewis and Clark overwintered with the Mandan Indians in 1804–1805 — people who now live on the Fort Berthold Reservation and in various urban enclaves. A graying man in black leathers throttled his Harley in beside me, and rested there astride. Birds called in the cottonwood. He pushed up his dark glasses. "I never been here," he said. "All my life."

I knew his feelings. Fort Mandan is sacred in western American mythologies. The Lewis and Clark story is an American creation myth. Their descriptions of the great West became actual in the imagination of the nation. The story of scientific and psychic exploration in the lands of the interior implied to generations heading for the West that their new homeland could be of both practical and spiritual significance. Their lives would not be wasted amid hinterlands.

But the Missouri today is a dammed-up, sluggish lake. The bluffs where the Corps of Discovery camped have been beneath water for decades. For tourists, there's an imitation Mandan village to visit in a nearby cottonwood grove, but my man on the Harley didn't give it a glance. "Must have been something," he said, as he dried his shades. Then he cranked up his machine, and swirled away into the mists.

By late afternoon, touring southward, I was in Strasburg, North Dakota, Lawrence Welk's hometown, a clean and meticulous, ordered settlement where I felt like I was in a movie set from the 1930s. Banks of yellow and lavender flowers were in full bloom and the lawns were freshly mowed. I parked and walked the sidewalks. The few fellow walkers I encountered studied me from the

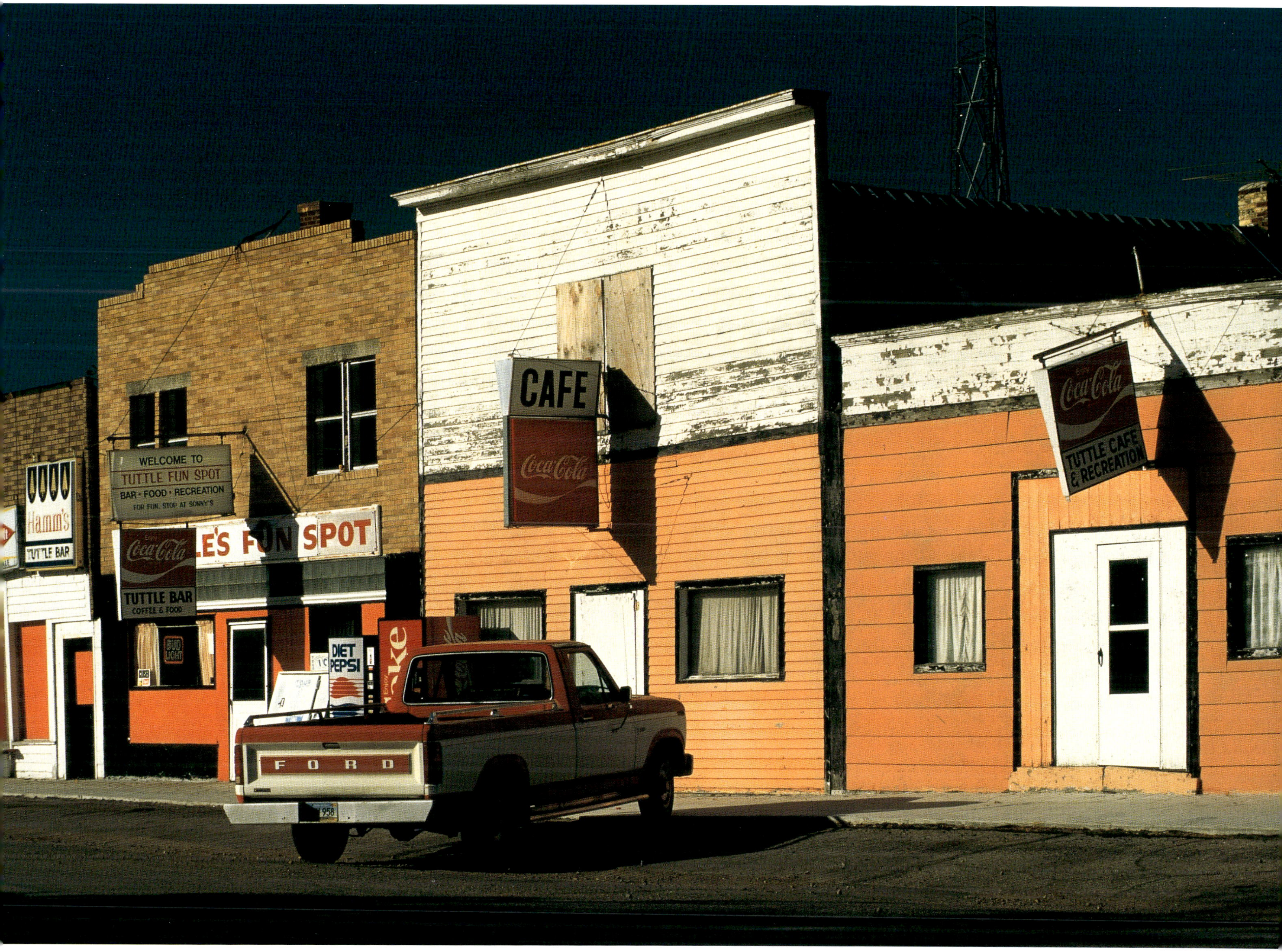

"FUN SPOT"
Tuttle, Kidder County, North Dakota

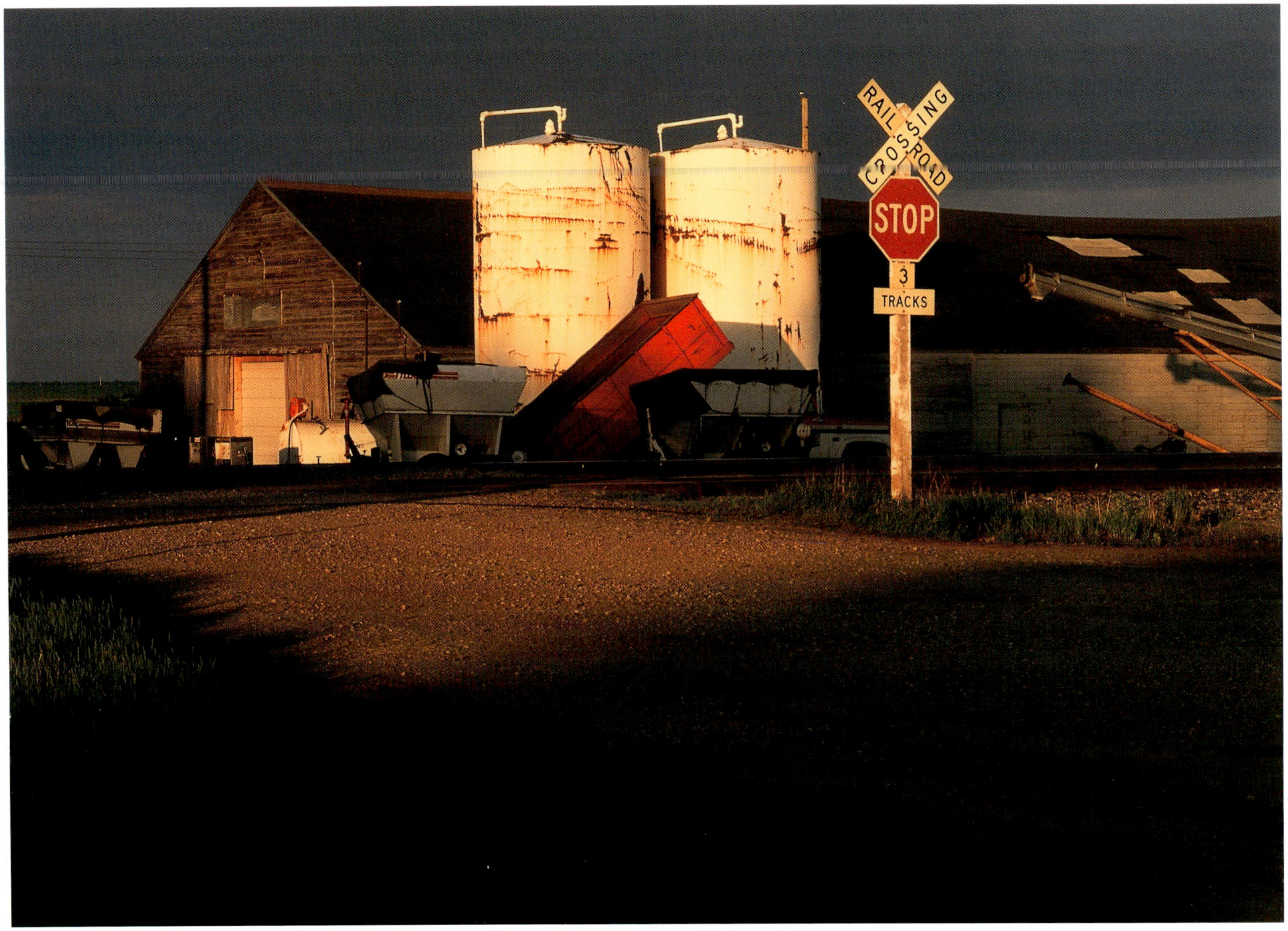

RAILROAD CROSSING
Selz, Pierce County, North Dakota

corners of their eyes and looked quickly away. I was a sixty-year-old stranger, nothing unique about my appearance, but a stranger, and thus, I think, suspect, an intruder, perhaps even dangerous, in some mysterious way capable of fracturing the quiet, careful dream of sufficiency being enacted in Strasburg. People there had put emotional barricades against encroachment from a barbaric outside world. Or so I thought. Strasburg felt to me like an emotional fortress, insulated and contained. Those feelings may have simply been a result of my insufficiencies. No doubt many people in Strasburg are as confounded and perplexed by the purposes of existence as anyone, and as easily, spontaneously generous.

But I kept wishing Judy Garland would come along in a pinafore, and lead people out to dance and sing and promenade on the wide streets, plucking flowers from the gardens, scattering petals behind. Crazy idea. Maybe that was the problem. Maybe Strasburg simply looked to be too walled-in, too incapable, or frightened and at the same time smug, to enjoy craziness. Surrounded by the enormous amplitude of that countryside, I felt loss, like a certain ration of possibility was gone from life in that vicinity.

So many little towns, like Mound City, down Highway 83 into South Dakota, where the average age of citizens seemed to be over seventy, are withering. Business is being done somewhere else. I yearned for gaiety and for clamor, an edge of uproar.

What I found, farther down the meridian, was Mobridge, a sport fishing town on the eastern bank of Lake Oahe, the 250-mile impoundment of the Missouri

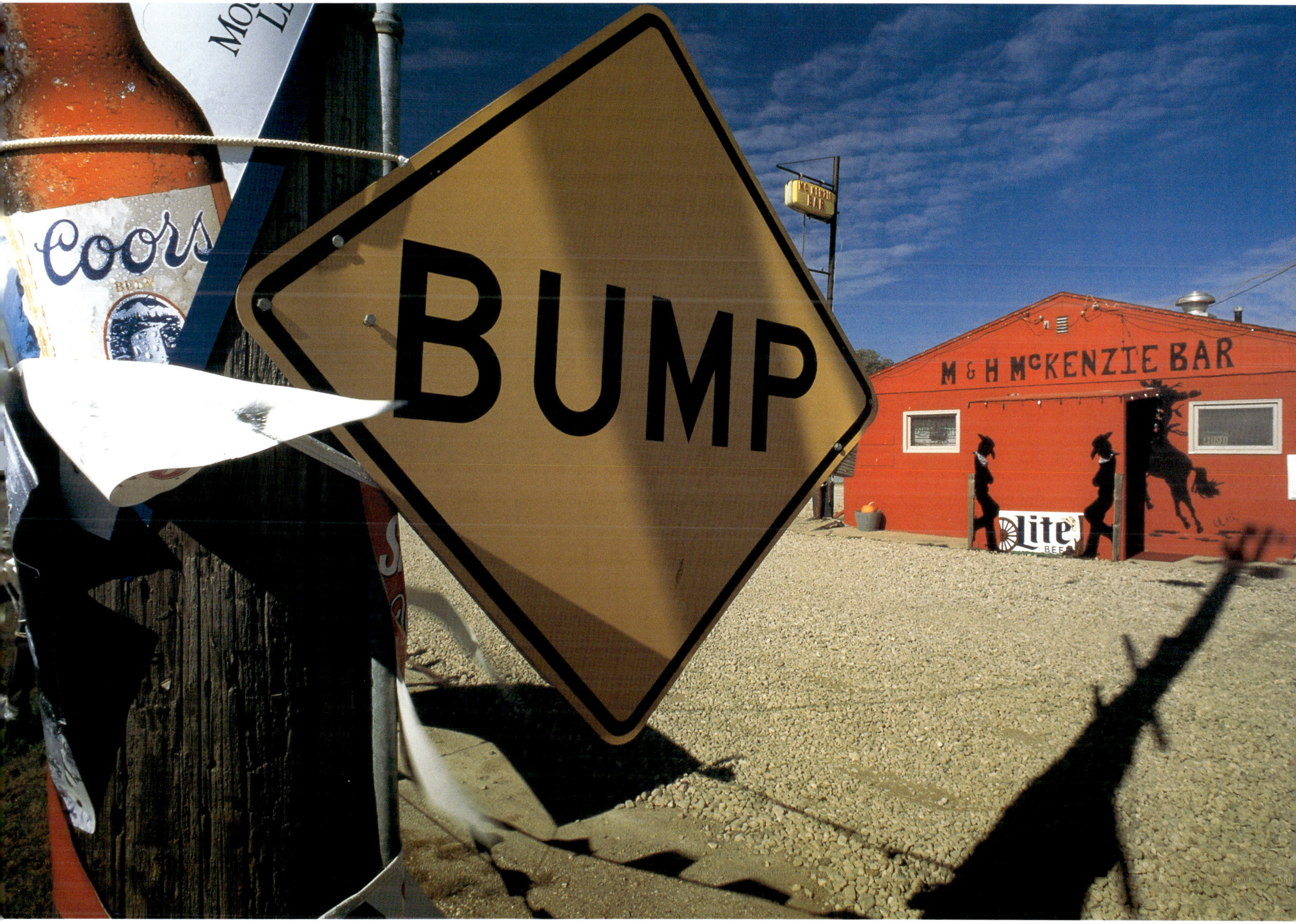

THE MCKENZIE BAR
Burleigh County, North Dakota

N O R T H   D A K O T A

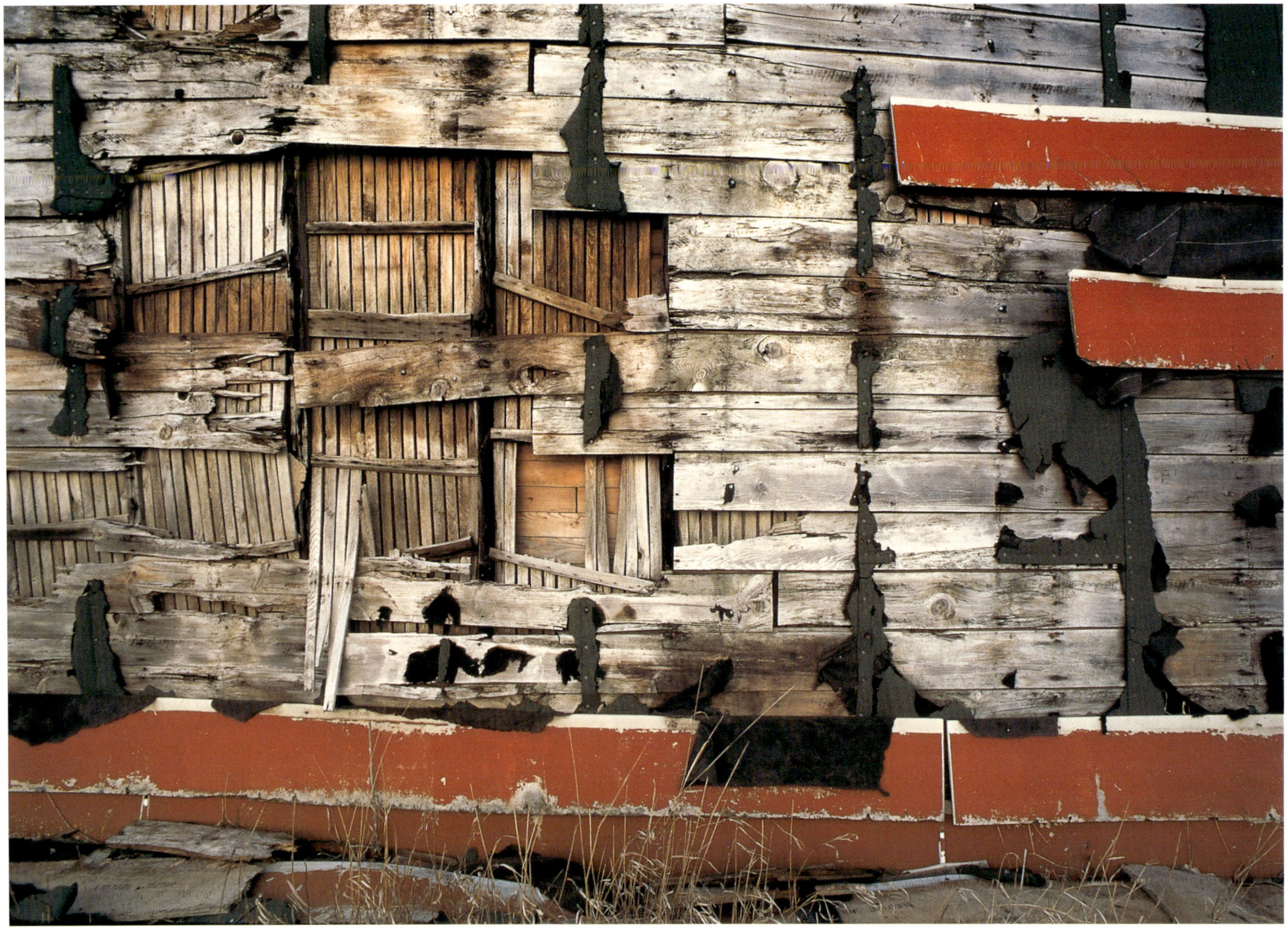

WEATHERED WALL
Esmond, Benson County, North Dakota

River that reaches from Pierre upstream past Bismarck and the site of old Fort Mandan. The Oahe Dam, built between 1948 and 1964, is the largest rolled earth dam in the world, 242 feet high, one of more than a hundred dams the Bureau of Reclamation built to generate cheap electricity, to store water for irrigation, and to control flooding on the Missouri while inundating some 600,000 acres of riverside agricultural land. Oahe Dam put one hundred and fifty thousand acres of farm lands in the Standing Rock and Cheyenne River Indian Reservations under water in the 1960s: a third of the reservation families were relocated; their abilities to sustain themselves were devastated; archeological and historic sites were drowned. We expend our treasures in profligate ways.

People come to Mobridge from hundreds of miles out on the prairies. Fishing, during the couple of decades after the dam was finished, was spectacular for walleye and smallmouth bass. Salmon were introduced in the 1980s. But lately, even from big boats, the catch is not what it used to be.

A cluster of good old boys were drinking and telling jokes at the rail in the lounge attached to the finest motel in town. I was happy to join them. But Mobridge is a declining tourist town, and gin and joviality are unconvincing cures for a pervasive sense that life's adventures have irrevocably run thin. The reservoir is silting full. In the bar we were fighting off the dreaded retreat to our motel rooms, television, and the silence of the digital clock.

■

The Missouri is an edge between ecologies.

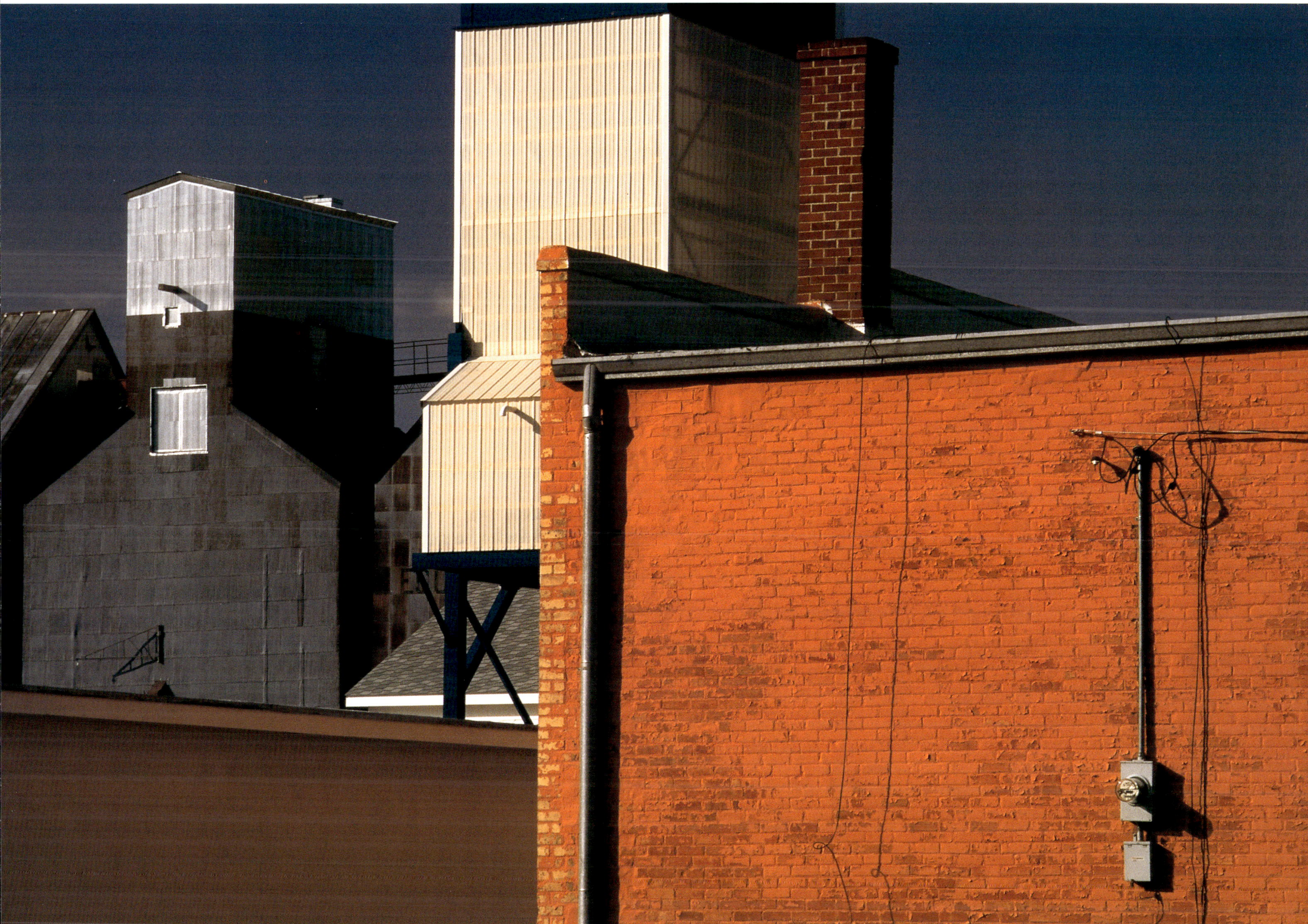

COMPOSITION IN ORANGE AND BLUE
Esmond, Benson County, North Dakota

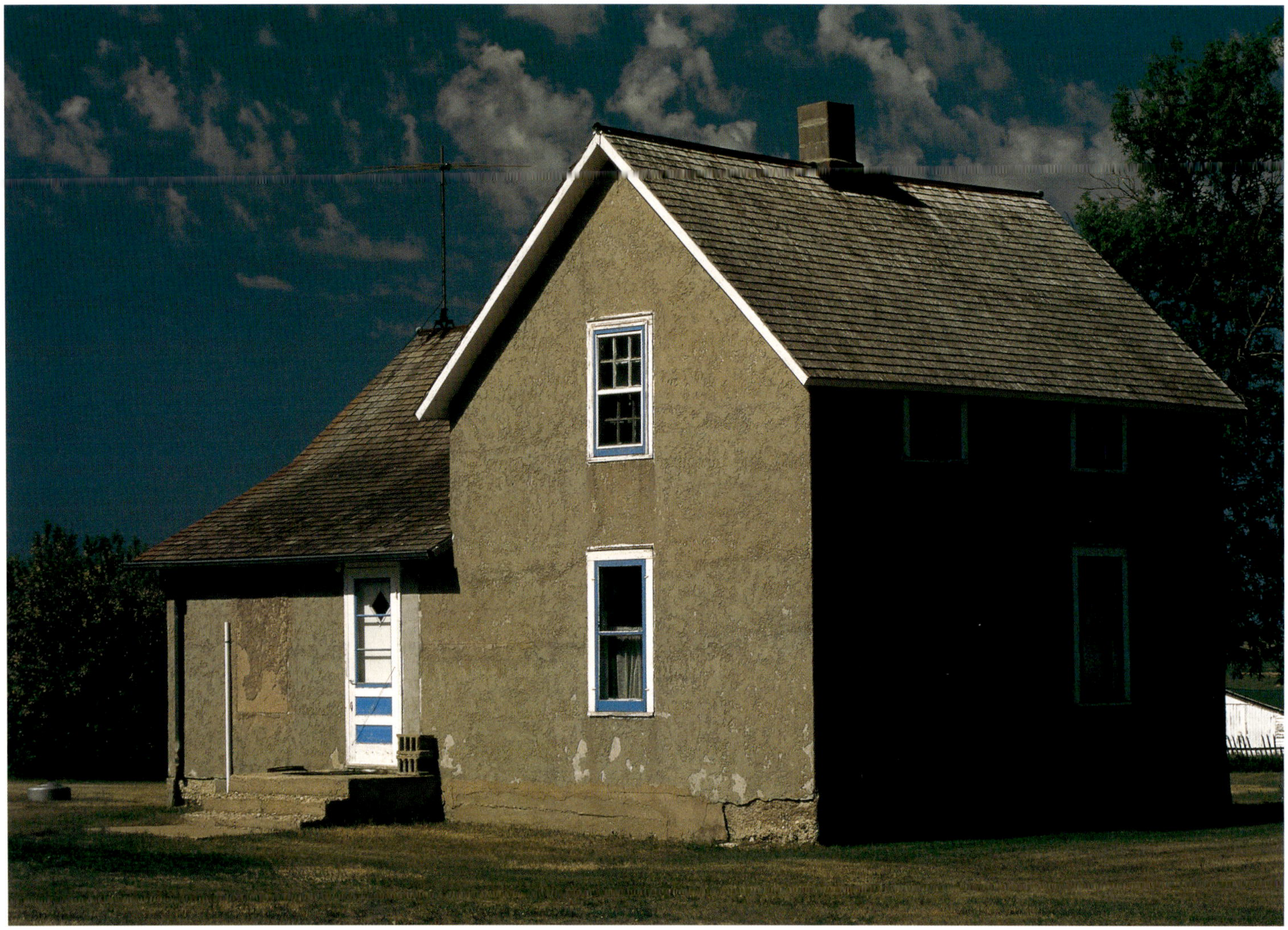

GRAY HOUSE
Tuttle, Kidder County, North Dakota

Precisely cultivated four-square fields reach endlessly off to the east; to the west are the unplowed grazing-land distances on the Standing Rock Indian Reservation.

The great leaders of the Lakota Sioux during the terrible years of their destruction as an independent people, in the 1870s and 1880s, were Crazy Horse, a mystic horseback prince of purity and vision, and Sitting Bull, a practical man dedicated to the survival of his people as Lakota, not as surrogate white men.

A road crew was repairing the long bridge across the Missouri. An Indian girl was flagging traffic. I asked about the memorial on the bluffs west of the river, commemorating the burial of Sitting Bull after he was killed while under arrest by Indian police the morning of December 15, 1890.

"You can see him," she said, and I could, I could see his statue far away against the morning sky. "They've shot off his nose," she said, and she smiled as if to say she knew that I was not complicit in such sacrilege. "Where do you live?" I asked. "Up the river," she said, and her hand, as she gestured, was crudely tattooed. "It's not anyplace," she said. Actuality was elsewhere.

Ours is a hunting species, and it's said humans are most comfortable when located on prospects, highlands looking out to the grazing plains, with escarpments, defenses, at our backs. Maybe that's part of the reason Westerners, used to the advantage of vistas, river valleys below and mountain ranges in the distance, have such a hard time focusing their vision on the prairies.

Undulating grasslands ran west to the

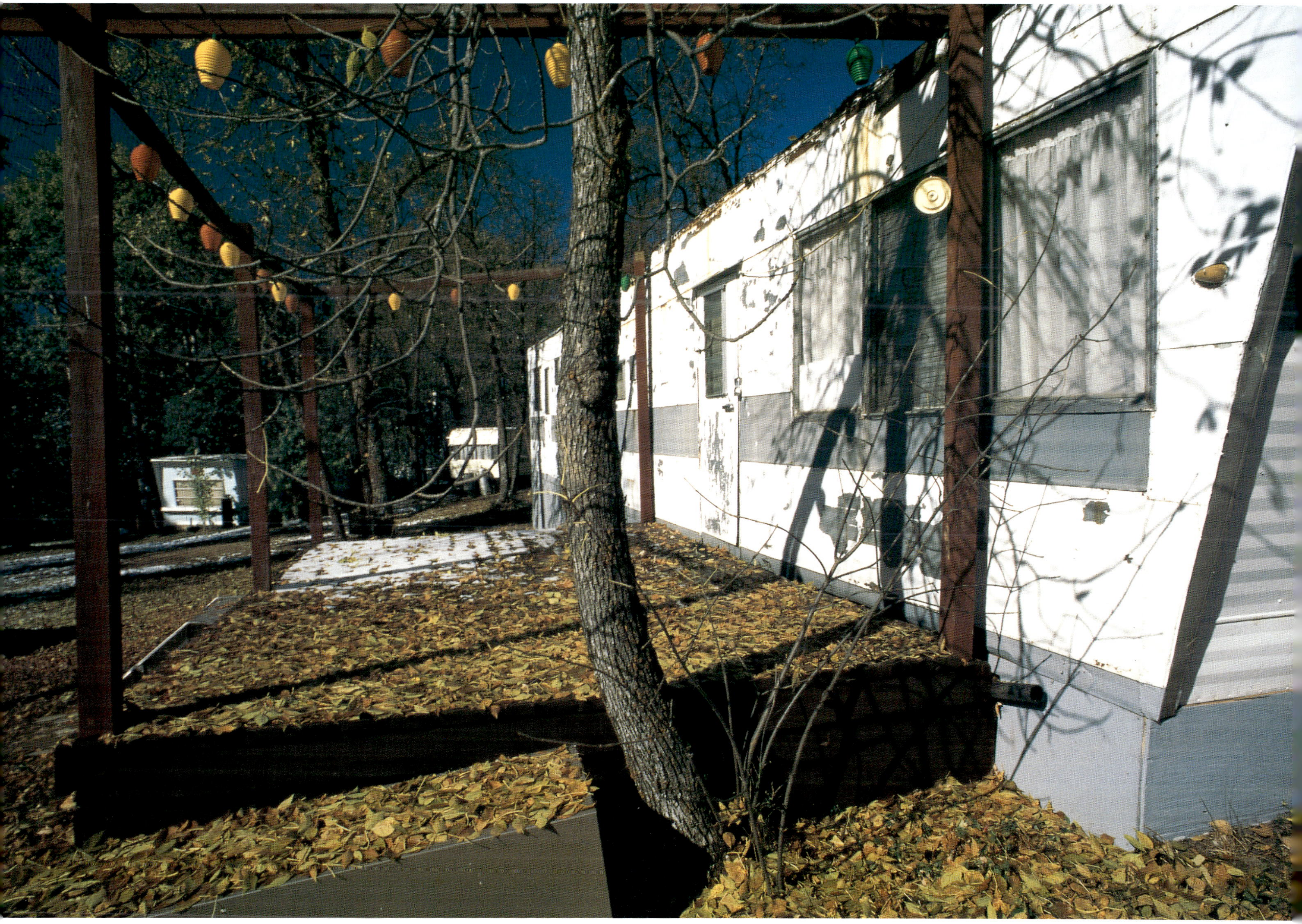

OFF-SEASON
Lake Metigoshe, Bottineau County, North Dakota

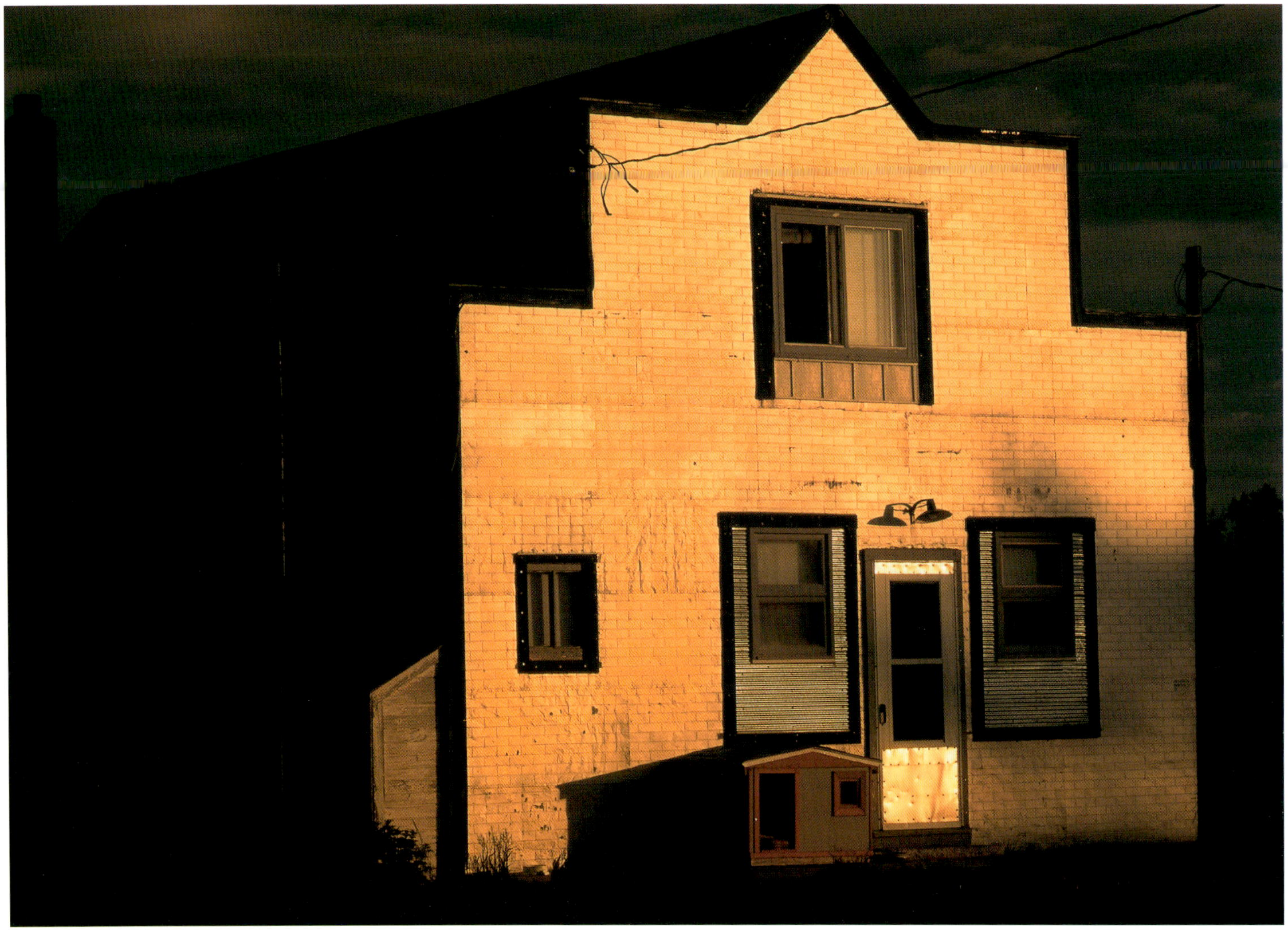

GOLD HOUSE
Selz, Pierce County, North Dakota

horizon form the Sitting Bull Memorial (his nose had been repaired). I stood with my back to Mobridge and the impounded river. The lights of an Indian casino shone on the highlands. What I saw were the trajectories of history.

By noon I was in Pierre, in the South Dakota Cultural Heritage Center Museum, which in seeming imitation of sod houses was burrowed into a windy hillslope. Manicured lawns swept down toward state buildings and the Missouri. I wandered in a $3 million display funded entirely by private and corporate monies, called "The South Dakota Experience" — spear points and wagons and plows and tools, examples of barbed wire, tepees and dioramas, audio-visual presentations, photographs, phonograph records — the past ordered for sanctioned reuse, scrubbed-up and virtual.

All down the 100th meridian, in the shrinking towns, there are storefront museums of pioneer life, collections of old dolls and antique shaving mugs. Alongside are lots filled with ordered rows of rusting farm equipment, harrows and plows, steam tractors from the homesteader days, tools of the fathers who invented this civilization. Preservation is a way of insisting on a specific past, saying these things count, a way of keeping faith, locating the self.

The nice woman in the bookstore warned me about weather. A tornado had that night wrecked a small Dakota town, and citizens were uncertain as to whether they wanted to rebuild or not. Maybe they would just give up, and move on, maybe life on the plains was too difficult after all.

"Keep an eye out," she said, having

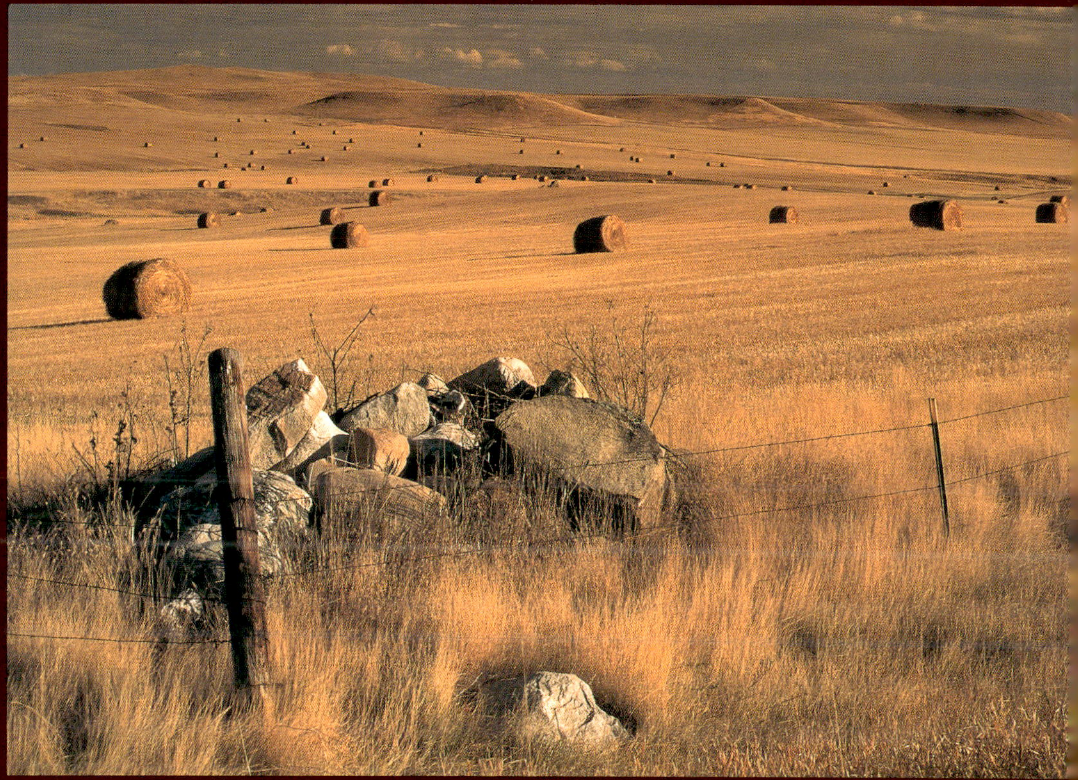

S O U T H   D A K O T A

ROCKPILE

East of Pollock, Campbell County, South Dakota

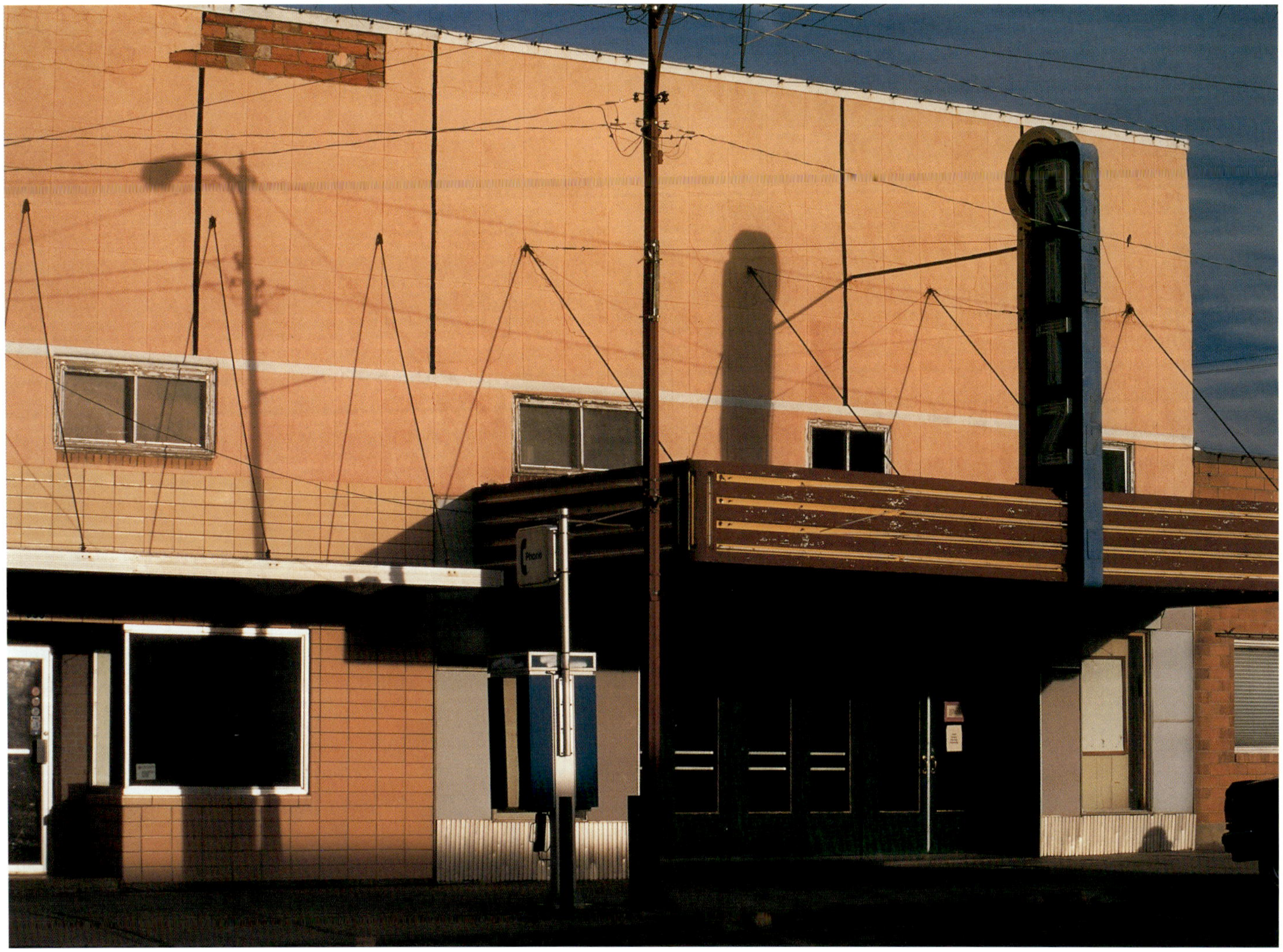

"THE RITZ" THEATER
Winner, Tripp County, South Dakota

turned utterly serious. "Those storms are terrible. You don't want to know." The enormous and uncontrollable weather is only containable in a story about the demonic, which exists outside the boundaries of the enfranchising story inhabited by community.

The Fort Pierre National Grassland south of the Missouri was blossoming and green as I headed down Highway 1806 (commemorating the year Lewis and Clark returned downriver) toward the Lower Brule Indian Reservation. The road turned to gravel and I turned back. Indian reservations are nations in our midst, where the defeated were sent to live. I wasn't willing to meet the eyes of people who had been truly disenfranchised. What if my car died, and I found myself dependent on their charities? Was I willing to admit, even to myself, that I didn't want to stop and spend any effort toward helping them? Surely they would have helped me.

The rest of that afternoon I flew south, the hull of my automobile up and riding on air. I was invisible and responsible for no one's trouble, lost in the pleasure of my own traveling. I braked to avoid a mother pheasant and her string of chicks; I stopped to contemplate an elaborate tumbling-down empty house, a metaphor for lost empires. I rushed on, trying to make my peace with the great rolling storm-cloud skies and the lift and fall of horizons. Tomorrow, I thought, tomorrow I'll really look at it.

Some towns were thriving while others died. As the farms consolidated, and roads improved, survival in business came to depend on the practical intelligence of indi-

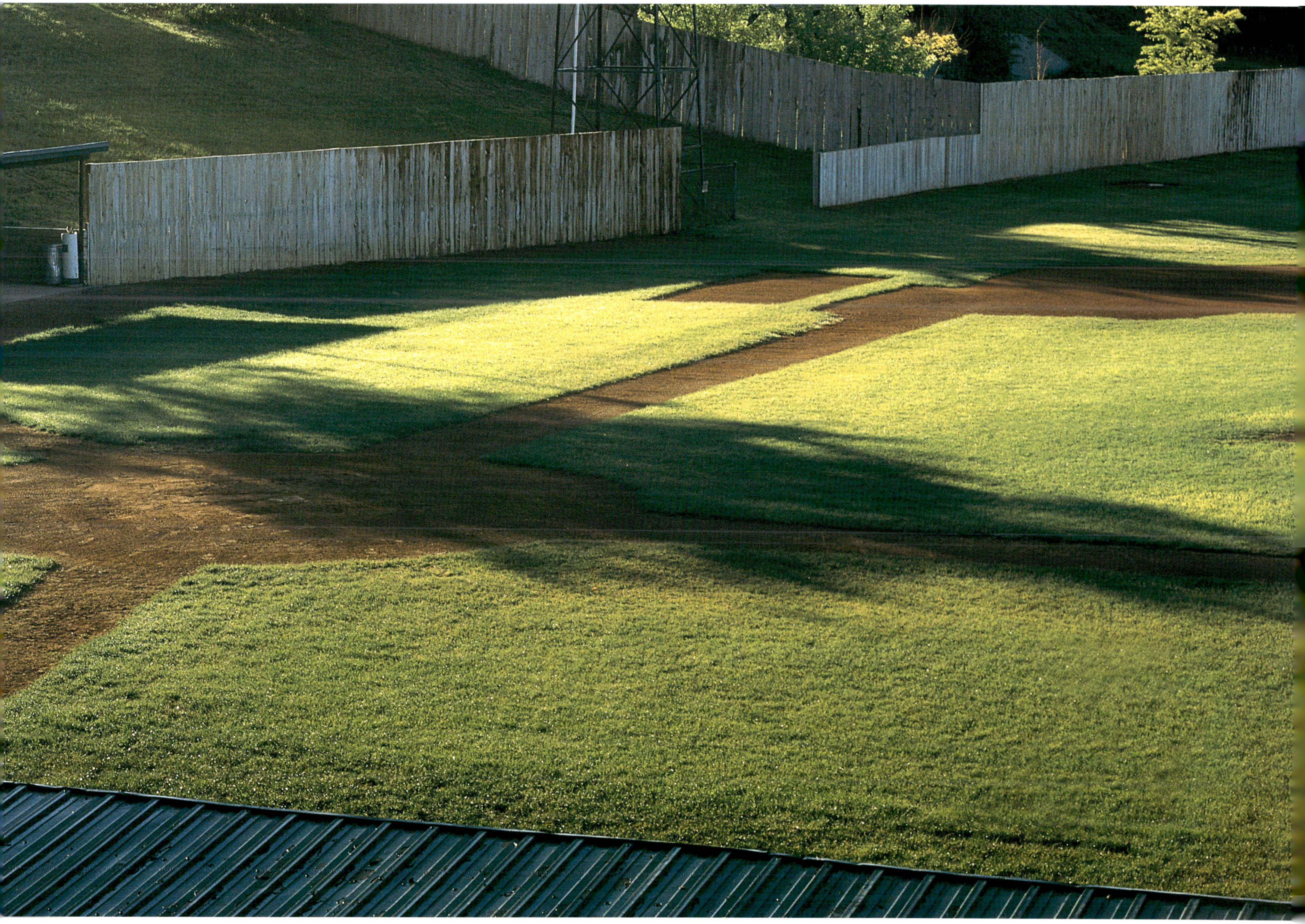

Diamond Twilight
Pierre, Hughes County, South Dakota

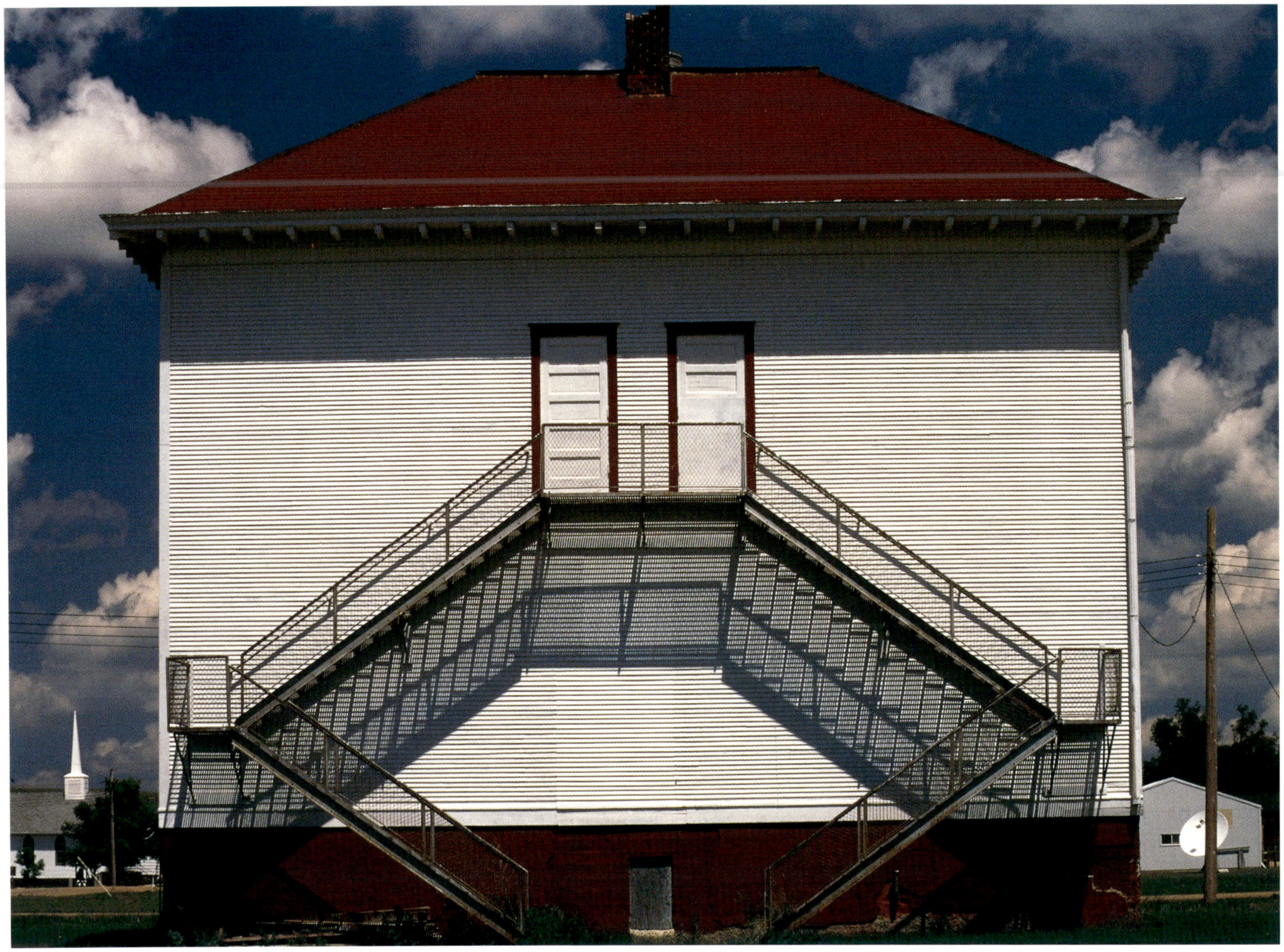

BEHIND THE SCHOOLHOUSE
Akaska, Walworth County, South Dakota

viduals, equipment dealers, and people who owned grain elevators. The survival of communities swings on the acumen of such people, and, also, on a clear-eyed community understanding that it's up to local citizens to care for what can be called their commons — doctors and nurses, churches and hospitals and old folks homes, teenage recreation centers and celebrations, rodeos and county fairs and yard sales and fifty-year wedding anniversaries with all the grandchildren present. Make your own list. The common denominator is taking care without lapsing into fearfulness. Winner, near the Nebraska border, lived up to its name. Bright green John Deere tractors and plows and combines were spectacularly displayed along the highway. Mechanization is another way of arming the self.

Winner, the "pheasant hunting capital of South Dakota," is the trading center for an agricultural kingdom, and becoming a little city. Franchise fast-food joints abound. I found this as discomforting as McDonald's in Madrid. I was visiting an antique world; I wanted it to stay rustic, and fit my preconceptions. This was another way of not seeing, or responding.

A towering ledge of black storm was forming as I drove the narrow highway at escape velocity. Until, passing the hamlet of Okreek, on the Rosebud Indian Reservation, I saw the surging wind had fallen quiet, the trees stilling themselves. Indian children were running excited routes in deep grass, and swinging reckless arcs on an old playground swing set. I had the sense to stop rushing past, and take some time to just watch their joy. This was something I could

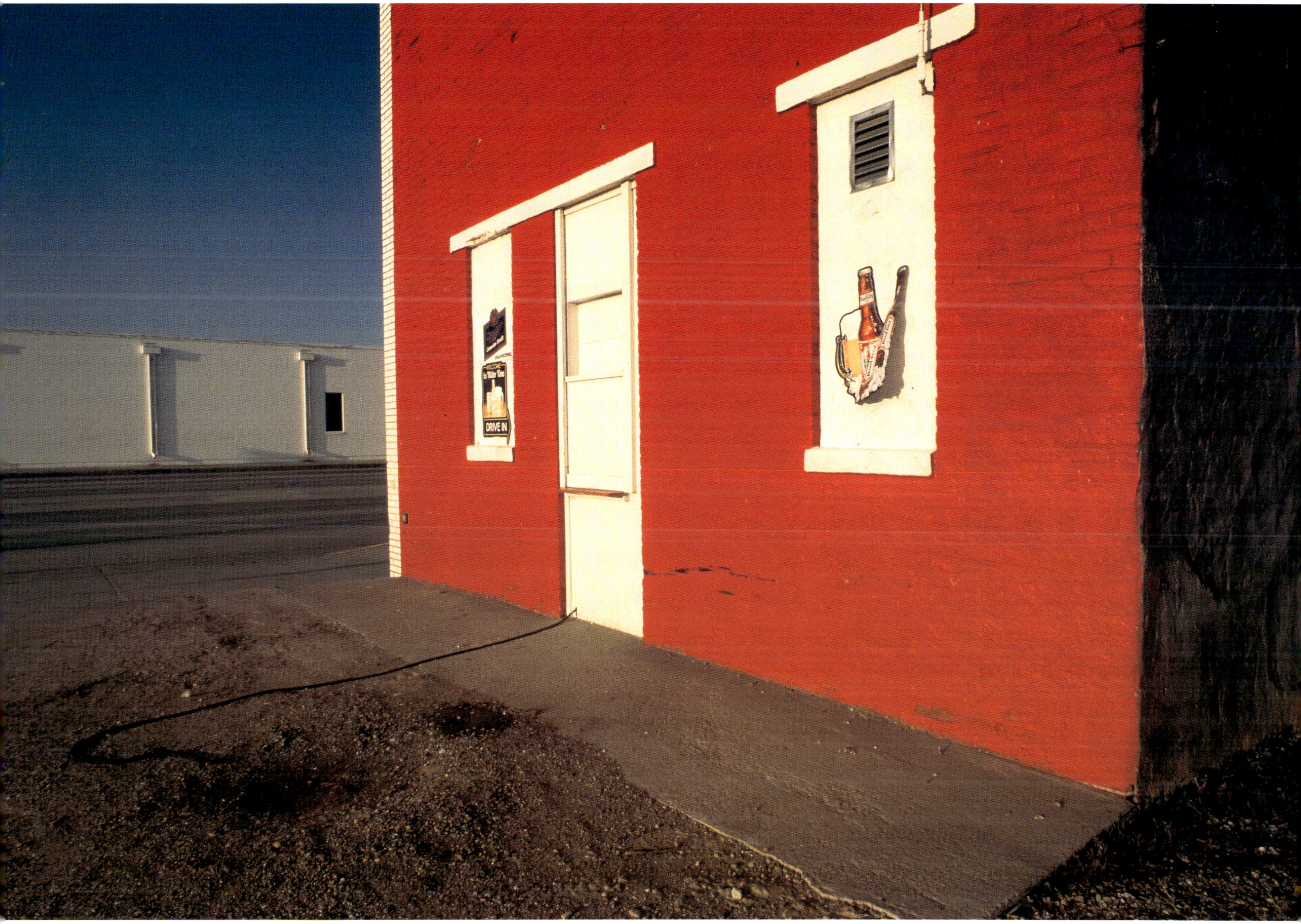

RED FRONT
Winner, Tripp County, South Dakota

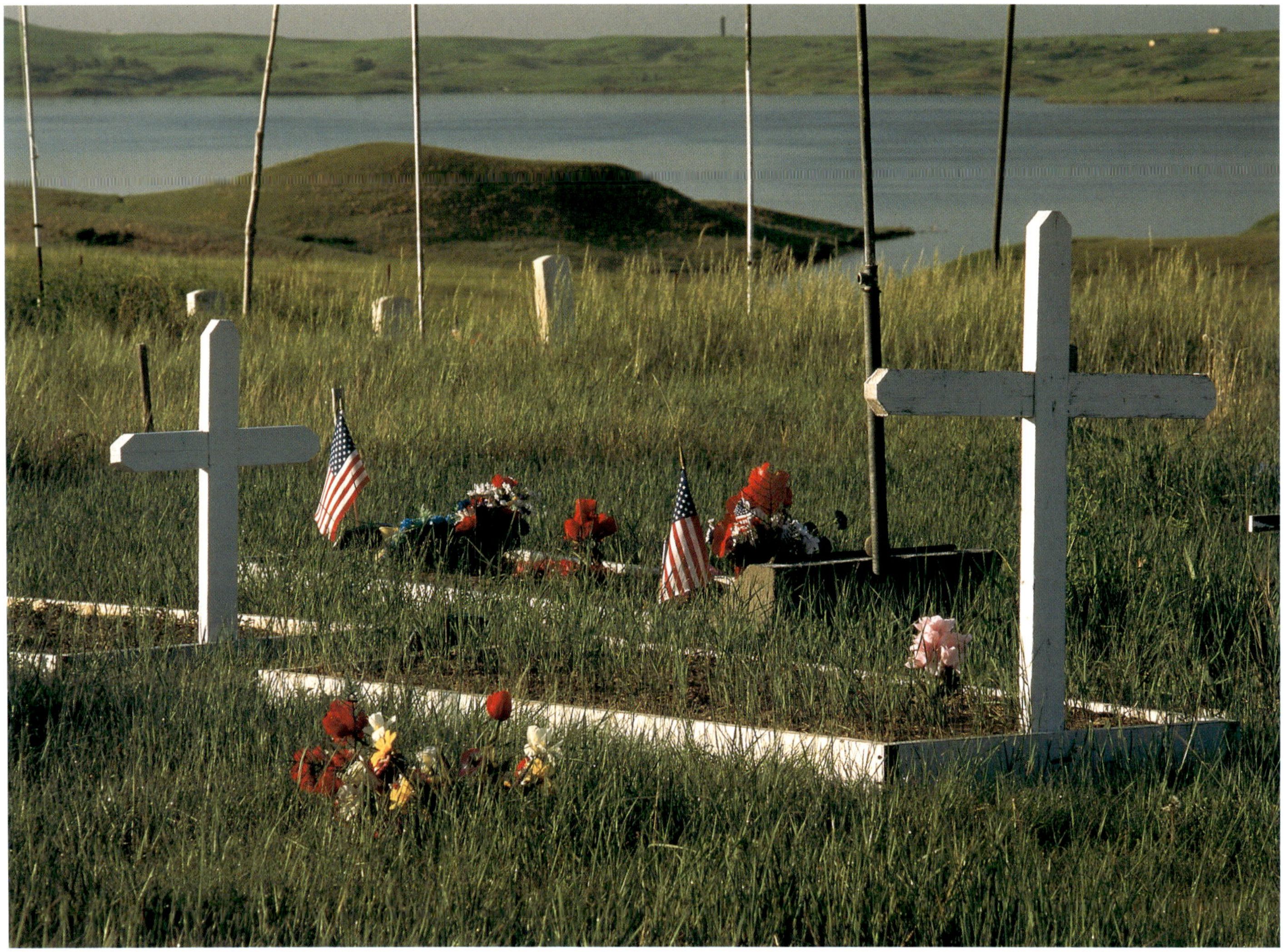

CEMETERY
Standing Rock Indian Reservation, South Dakota

look at, the little boy in his lavender shirt, and his bright fearless smile.

The Rosebud Reservation, where the single white buffalo sacred to the Lakota once lived, and so long ago vanished, where the Lakota people live out the rest of time, is another remnant Indian homeland. It is also the fourth poorest county in the United States. The little town of Mission is site of the Sioux institution of higher education, Sinte Gleska University. A very dedicated young white woman I know, who taught there, says her work was constantly against a strain of despair and anomie that runs with poverty inside wrecked and desperate cultures. "Try it," she said, "teaching hope." She's moved on to a job at a major university in Arizona. "Rosebud," she said, "was too difficult for me."

The Plains are haunted by ghosts and by history, stunning stories of unthinkable injustice and heedlessness. So much unfairness, so close to home, its sources ardently ignored. The history of what happened during the decades of conquest and settlement is still, for many there, emotionally untouchable. A pair of wild-haired young Indian hitchhikers, a boy and a girl so far as I could tell, were heading south from the edge of town. I didn't consider picking them up. I wondered if a white farmer, earning a living on land taken from the tribes, would be any more trusting, and thus generous, than I was.

At the Nebraska border, outlined in neon, a long-stemmed rose glowed in the twilight like an apparition on the wall of an Indian casino called the Rosebud. The interior was windowless as a cavern. Hunched

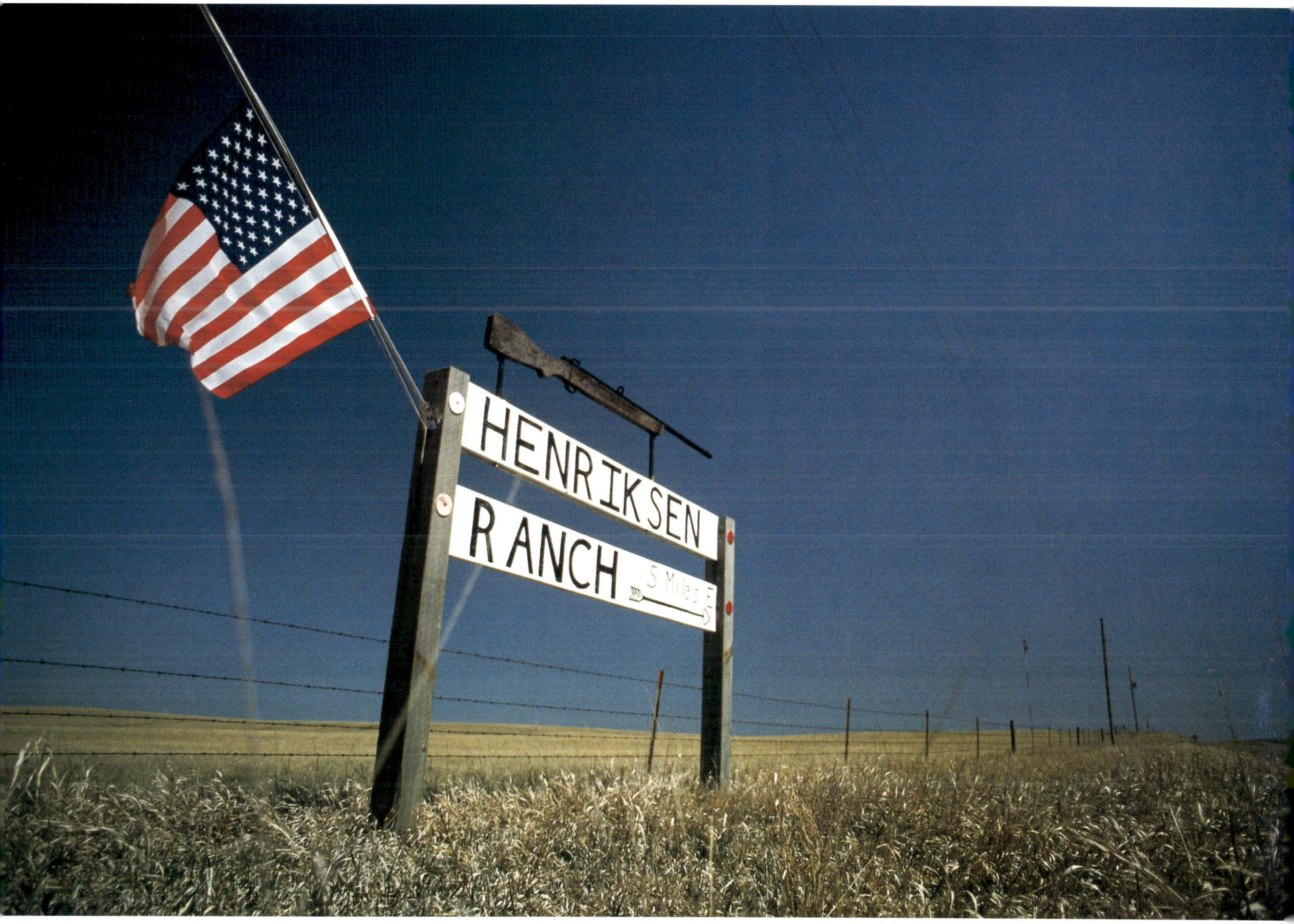

HONORING THE GOVERNOR
Highway 83, Lyman County, South Dakota

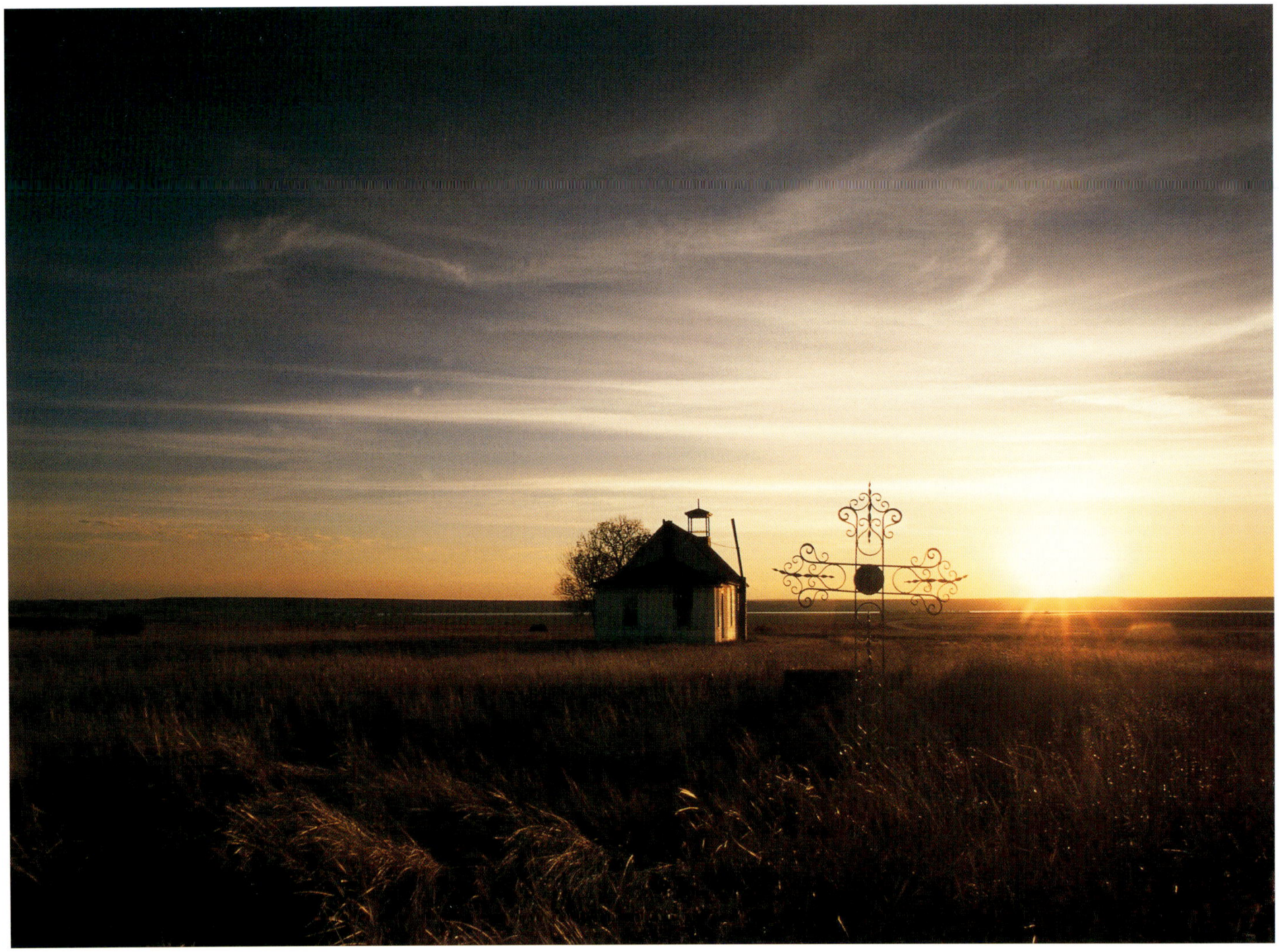

EMPTY CHURCH
Lower Brulé Indian Reservation, South Dakota

over an electronic blackjack machine, shoving quarters, I might have been for a moment part of the spangled Great World seen in television advertisements for Las Vegas. When I left, the vast weather had gone east without me. The grasslands were still, birds calling.

Without thought, in twilight outside the Rosebud casino, I was thrilled by whatever happenstance luck had brought me to this. Connection to glories, which leads us to intuit that our lives cannot help but be meaningful, is a bright lure, and the fundamental reason why people are so willing to spend their lives in places like Rosebud. Or so I think. It is, however, private and difficult to articulate, a religious experience. The air resonated with significance and possibilities. This, then, might be the secret.

The Sandhills of Nebraska are an ancient bed of fossil dunes covering 19,300 square miles, reefs of grasslands broken by little lakes and pockets of marshland, cut by a few small rivers. It is a sort of secret paradise, at least for the old-time cattle ranching families who own most of it (you are said to be only at play in the cow business if you own less than 5,000 acres in the Sandhills). The backlands West has always been a safety valve, a country where overflow populations could settle. Established locals would like to see that notion die. They want to think their part of the West is theirs forever.

This town of Valentine, on the Niobrara River at the northern edge of the Sandhills, is uneasily preparing to enter the tourist-servicing, amenities-oriented "new" West. They have a paved bike path on an old rail bed through the center of town, and river

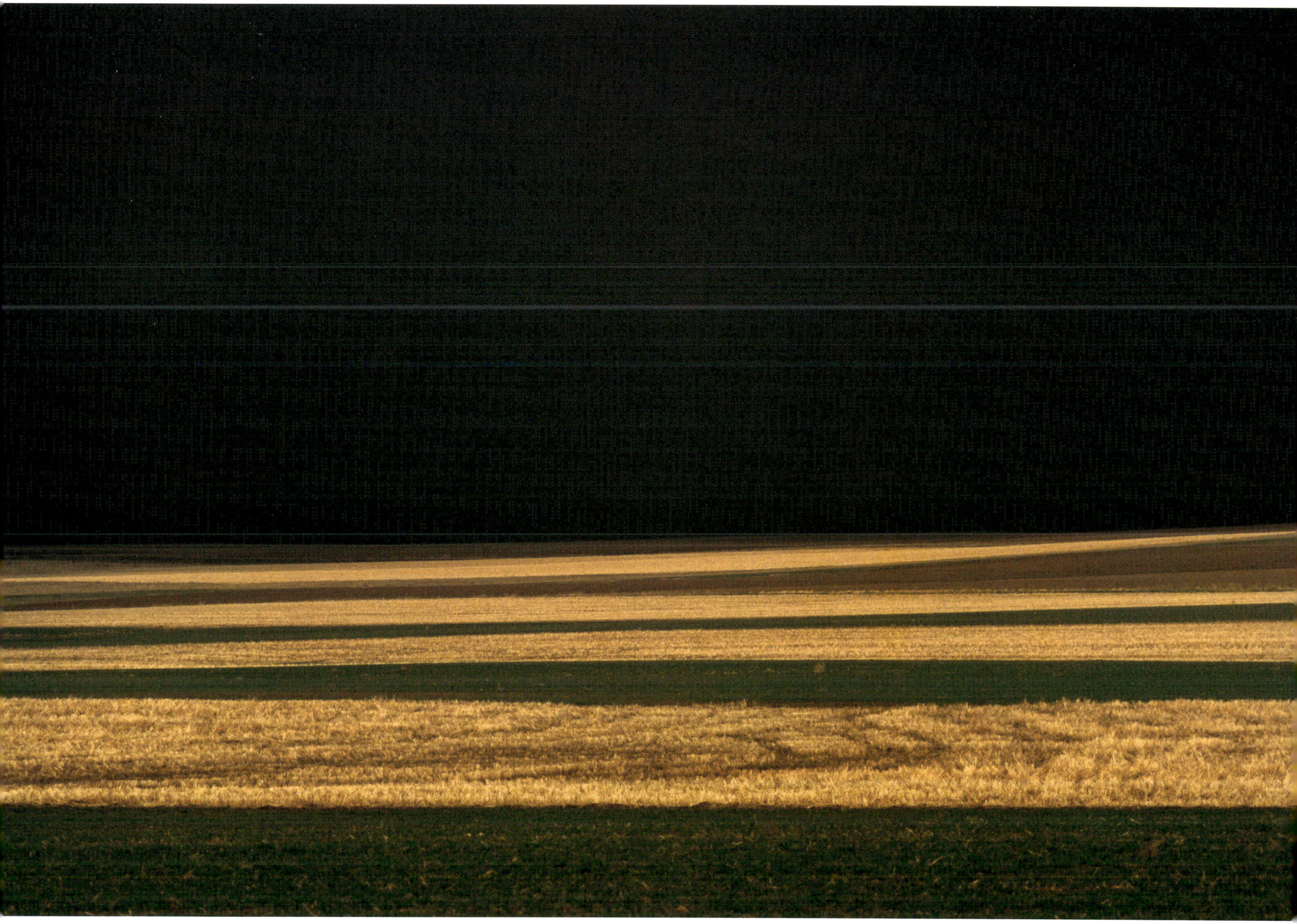

STORM SKY
South of Presho, Lyman County, South Dakota

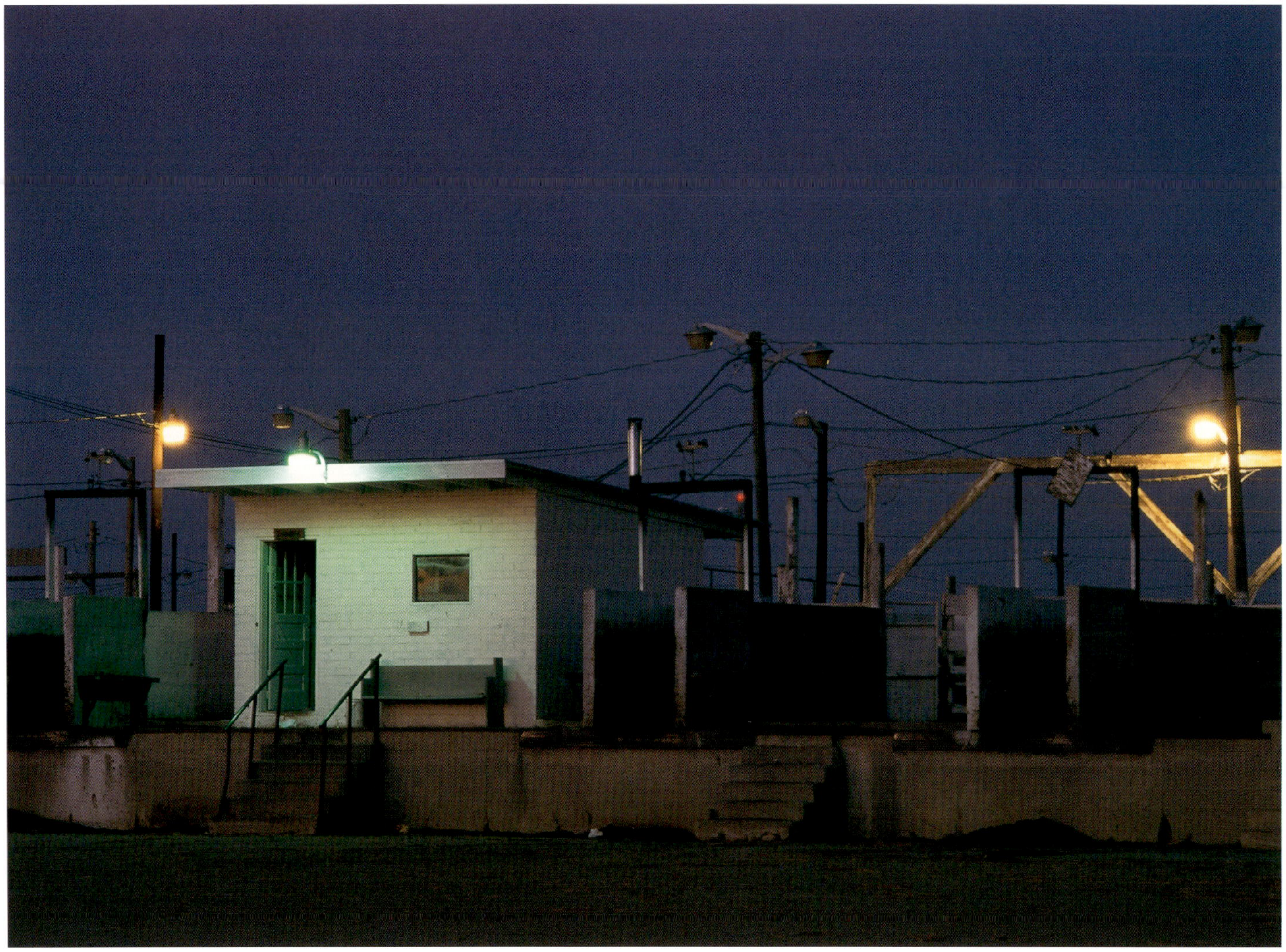

WINNER LIVESTOCK AUCTION CO.
Winner, Tripp County, South Dakota

rafting (the Niobrara has been designated a national "Wild and Scenic River") is being promoted alongside rodeos.

For the last century, in at least the cowhand West, ranchers exercised absolute economic hegemony. But now they no longer get to call all the shots. The merchants in Valentine can make a good living catering to tourists. But they can't sell the outdoors if the gates are locked. Another world is coming to the Sandhills. South of Valentine, near the town of Mullen, there's a high-end golf course designed by Masters Champion Ben Crenshaw to replicate seaside Scottish dune-land links. Here, golfers from all over the world come to play. They fly in, bringing money and their invulnerabilities to condos beside the course, and they don't much care about the prerogatives of ranchers. They play, and they fly out.

Duane Gudgel sells books about the plains out of a store called the High Plains Trading Company, all the best books about the plains. There would have to be at least a single great narrative about the settlement of the plains, and there is, *Old Jules*, written about her father, and life in the Sandhills, by Mari Sandoz. Gudgel also had everything by Willa Cather, and titles by Teddy Blue Abbott, John Graves, Wallace Stegner, and James Galvin. An evolving literature is a sign, we agreed, that the West is cooking up a culture that is capable of deciding what it wants to be.

"Times change," Duane Gudgel said. "White men have only been in these hills for a hundred years. There's not one thing written in stone." He's right, of course. Tourists are likely to be followed by retirees,

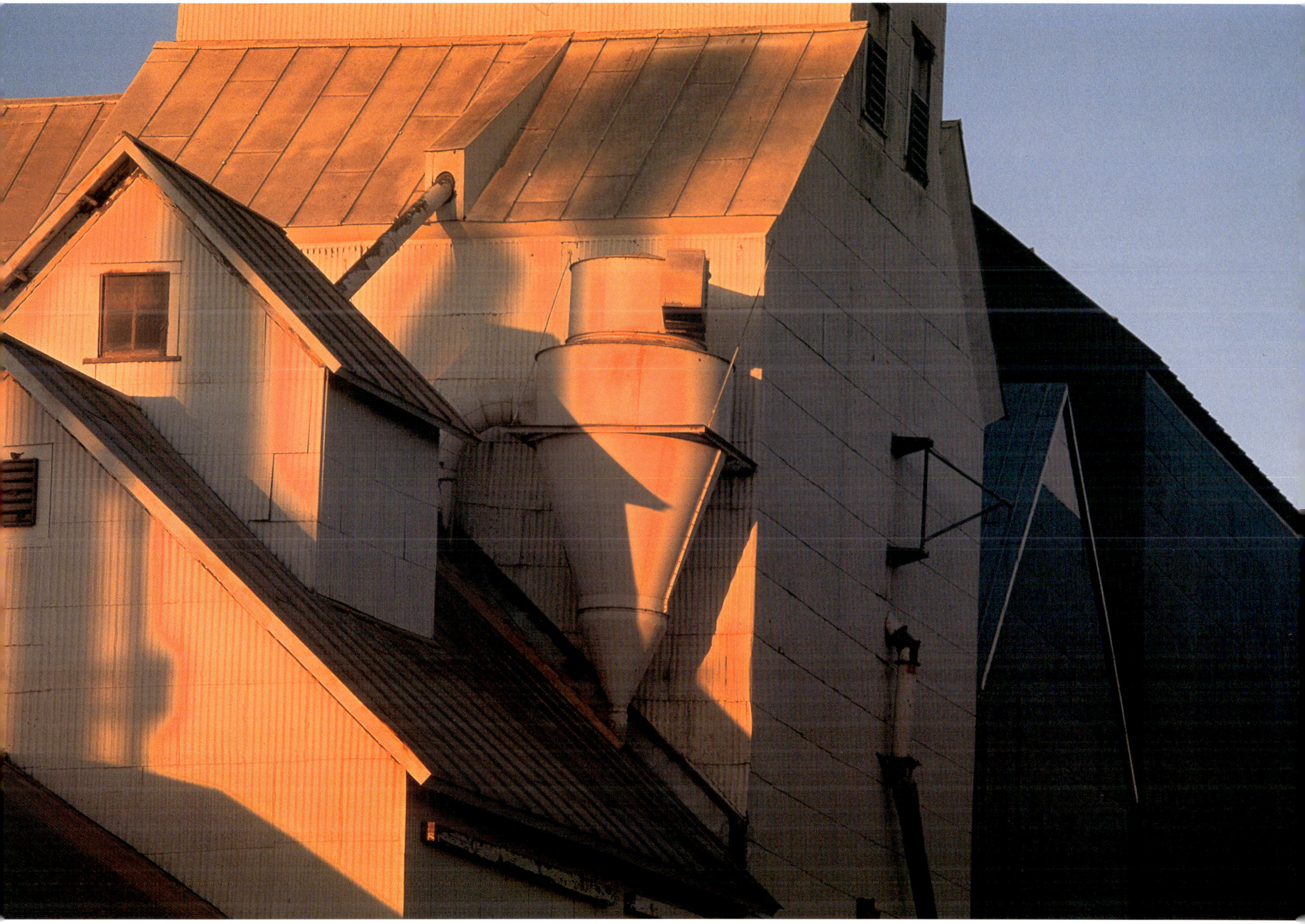

ELEVATOR SHADOWS
Gettysburg, Potter County, South Dakota

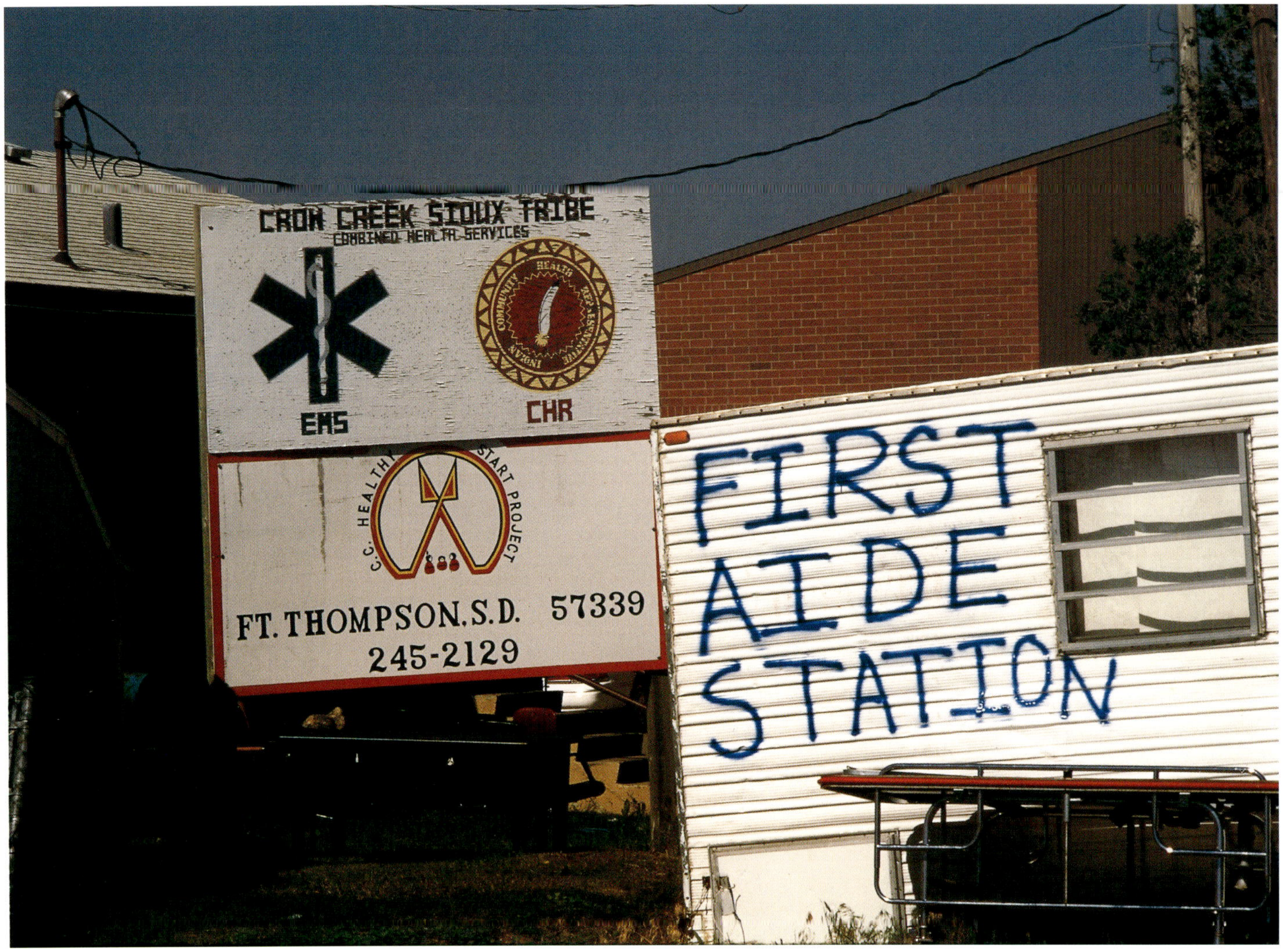

"FIRST AIDE"
Crow Creek Indian Reservation, South Dakota

who will settle in and vote, often with marked indifference to the established habits and customs of local citizens.

Ranchers in Montana, feeling crowded, are moving their operations to the Sandhills. Westerners in deep country torment themselves with the idea that the West faces a diminished future. Some claim that their defeats are like those suffered by the Indians. In that, privileged people risk sounding ridiculous.

By lunchtime I was at the southern edge of the Sandhills, in the town of Arnold, in a cafe where the girlish waitress and a sunburnt young man in paint-stained trousers were conducting a light-hearted investigative flirtation. "Twenty?" she said. "I never been out with anybody who was twenty."

Forking away at my macaroni and cheese, I was made happy by snooping on their beginnings. Sweet times are still abundantly possible.

■

Despite the very true stories we hear about economic despair (masses of farmers going broke, the populations still falling), farming along the 100th meridian looks to be hugely productive, and perfectly capable of generating prosperous, well-kept, and generous communities. These thoughts came to me while I passed through the Platte River country in southern Nebraska, an elegantly watered and farmed countryside, and little cities like Gothenburg — its immaculate parks sprinkled with children on their bikes in the sunlight.

The German-American community of Eustis was preparing for a day of festivity. Young men were raising a new tavern sign

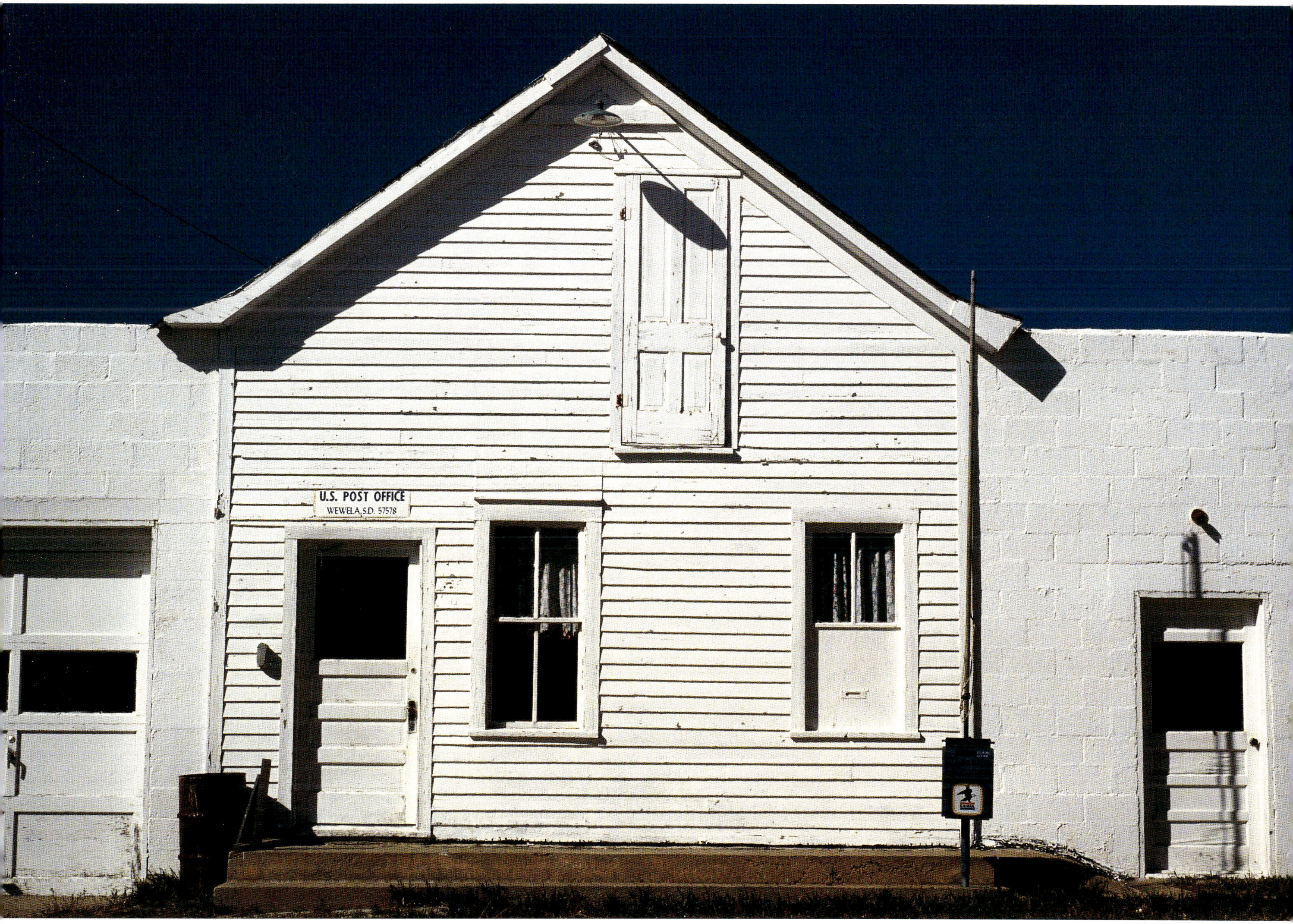

White Post Office
Wewela, Tripp County, South Dakota

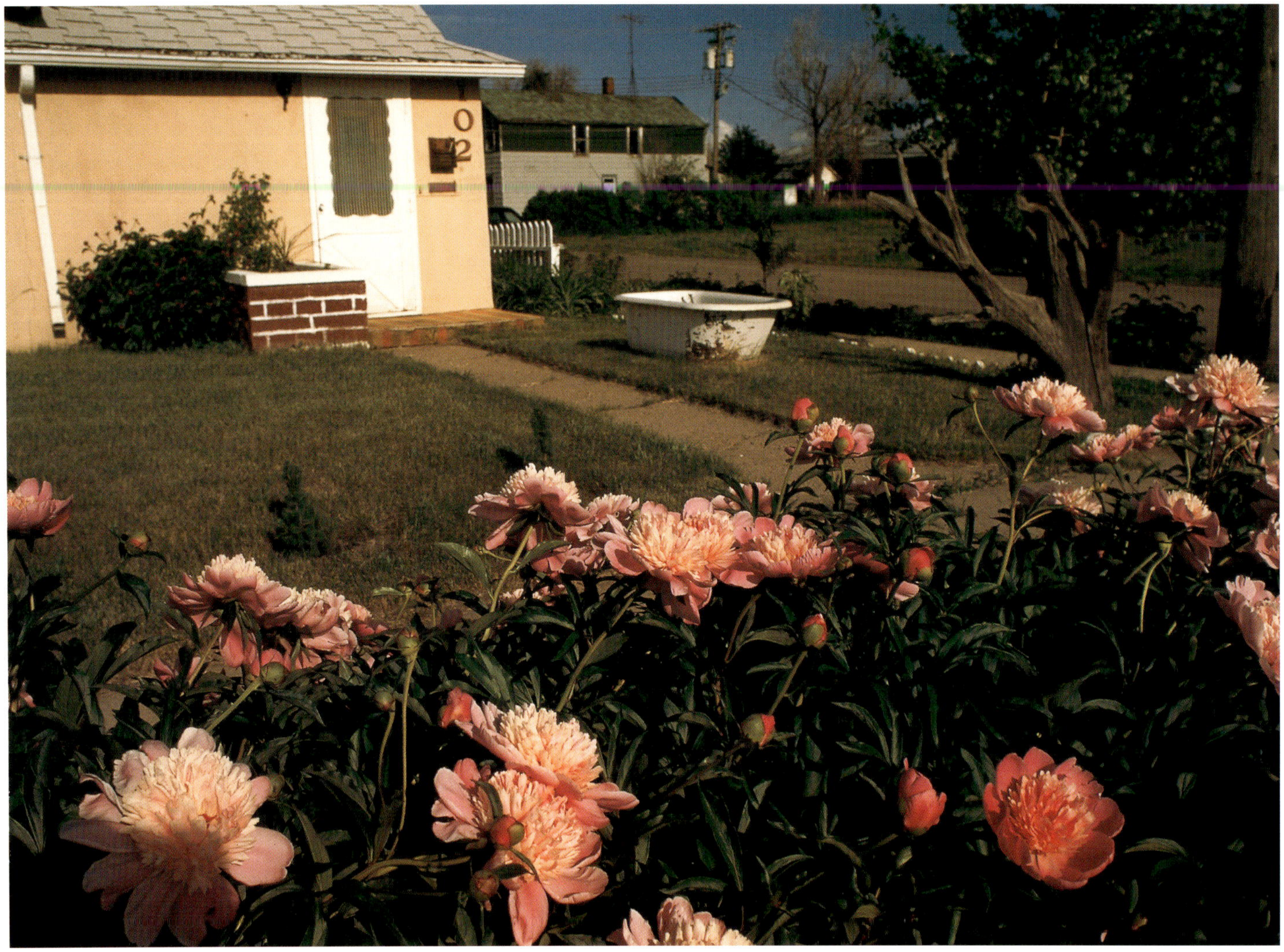

PEONIES
Mobridge, Walworth County, South Dakota

with an antique Farmall tractor. The shaded town and the clean farmlands were part of the same functional community. Enclaves of the kind that had seemed emotionally barricaded in the Dakotas now looked to be simply tranquil. I don't know why. Maybe I was getting comfortable in what I had first seen as alien territory; or maybe those little towns were actually more open-hearted, more welcoming. The countryside was elegantly farmed, contained by watercourses, all the way south to the Kansas border. Then it opened on vastness, wheatfield beyond wheatfield, towns marked by towering concrete grain elevators.

South of the Republican River, when the essentially shadeless, treeless, plains of western Kansas and Oklahoma and Texas opened before me, the southern sky and horizons expanded in a way that seemed to say, see, your existence is inevitably connected to infinities, and will never be exactly fathomable. Possibility loomed and contracted at the same time. As in the Dakotas, there was nowhere to hide.

During the latter years of the nineteenth century Mennonite German-Russian farmers were being forced out of the Volga River region of Russia for political reasons. Many came to the plains of Kansas, bringing Asian wheat seed with them, called Turkey Red. It thrived. Wheat from thousands of acres seeded to Turkey Red was being fed through steam-driven threshers in western Kansas by 1880.

During World War I a boom market for wheat drove farmers on the Plains to plow up more and more native grassland. In the 1920s a falling market drove them to plow

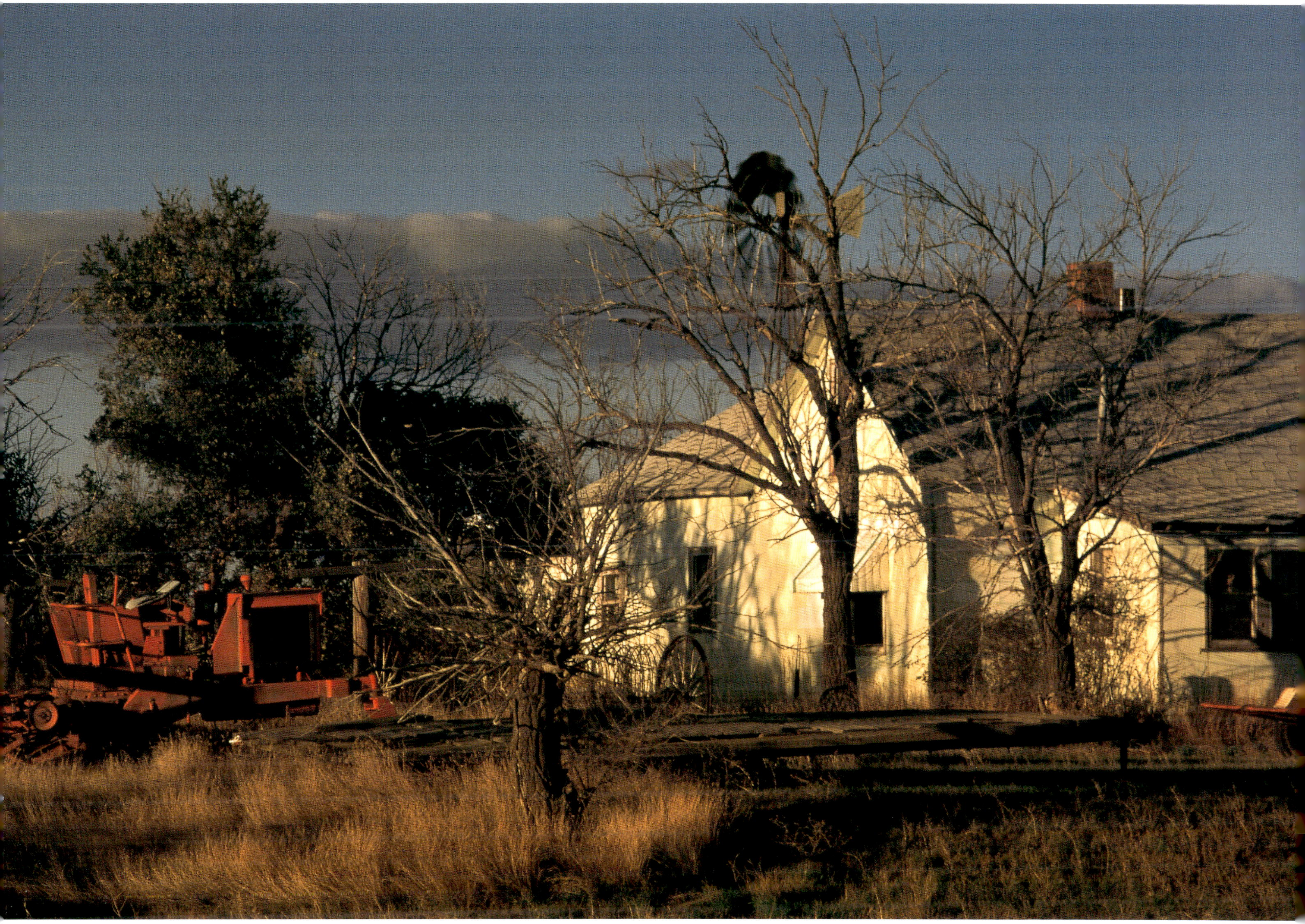

SOMEONE'S PLACE
West of Gettysburg, Potter County, South Dakota

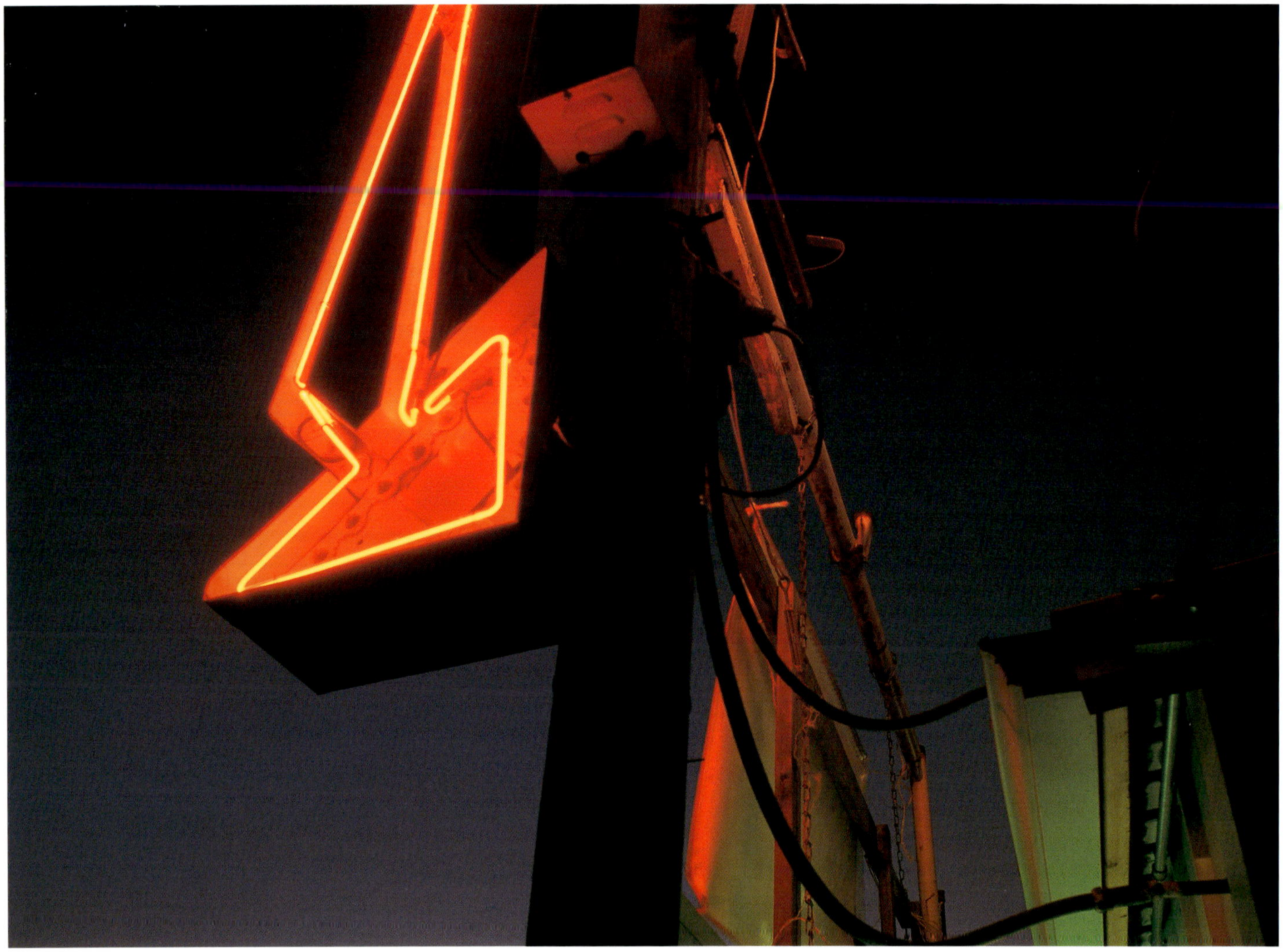

BOOZE HERE!
Winner, Tripp County, South Dakota

yet more ground. The answer to every problem was understood to be more production. Soils deteriorated. Drought came to the plains in 1931, and crops didn't germinate. Dry powdered soils began to blow.

The Dust Bowl was an economic and environmental disaster. The most famous of the dust storms came on what was called Black Sunday, April 14, 1935, swirling up out of western Kansas. Cattle suffocated and died; dust fell in New York, and on ships at sea. In 1936 Ernie Pyle visited and wrote, "If you would like to have your heart broken, just come out here. This is dust-storm country. It is the saddest land I have ever seen."

"As far as the eye could see," Pyle said, "there was nothing. There was not a tree, or a blade of grass, or a dog or a cow or a human being — nothing whatever, nothing at all but gray raw earth and a few far houses and barns, sticking up from the dark gray sea like white cattle skeletons on the desert. There was nobody in the houses. The humans had given up and gone. It was death, if I have ever seen death."

The nation paid farmers to not plant, in hopes of returning at least some of the Plains to grassland. But drought ended, and World War II drove up the price of wheat, and farmers inevitably went back to plowing. The small farms were sold and consolidated into agribusiness layouts. More stories were buried in the landscape. Machinery implacably replaced tenant farmers and laborers.

These days, a lot of the crops in western Kansas are irrigated with deep-well water from the enormous Ogallala Aquifer. But

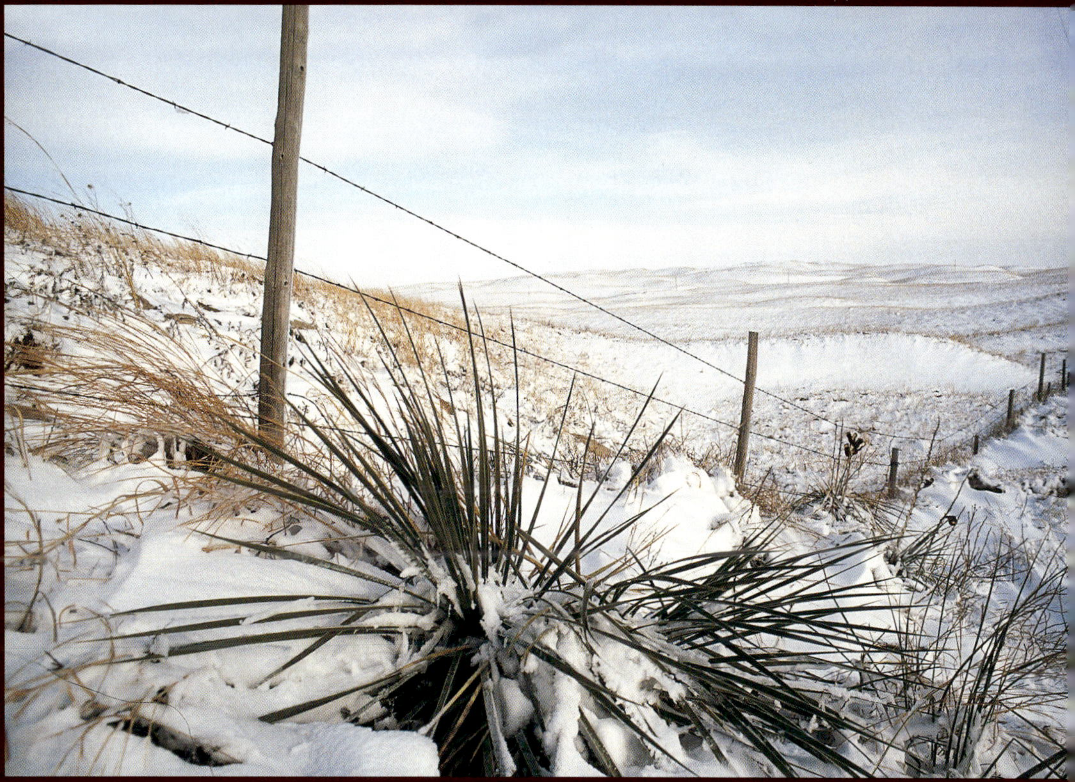

N E B R A S K A

SNOWY FENCE
North of Thedford, Thomas County, Nebraska

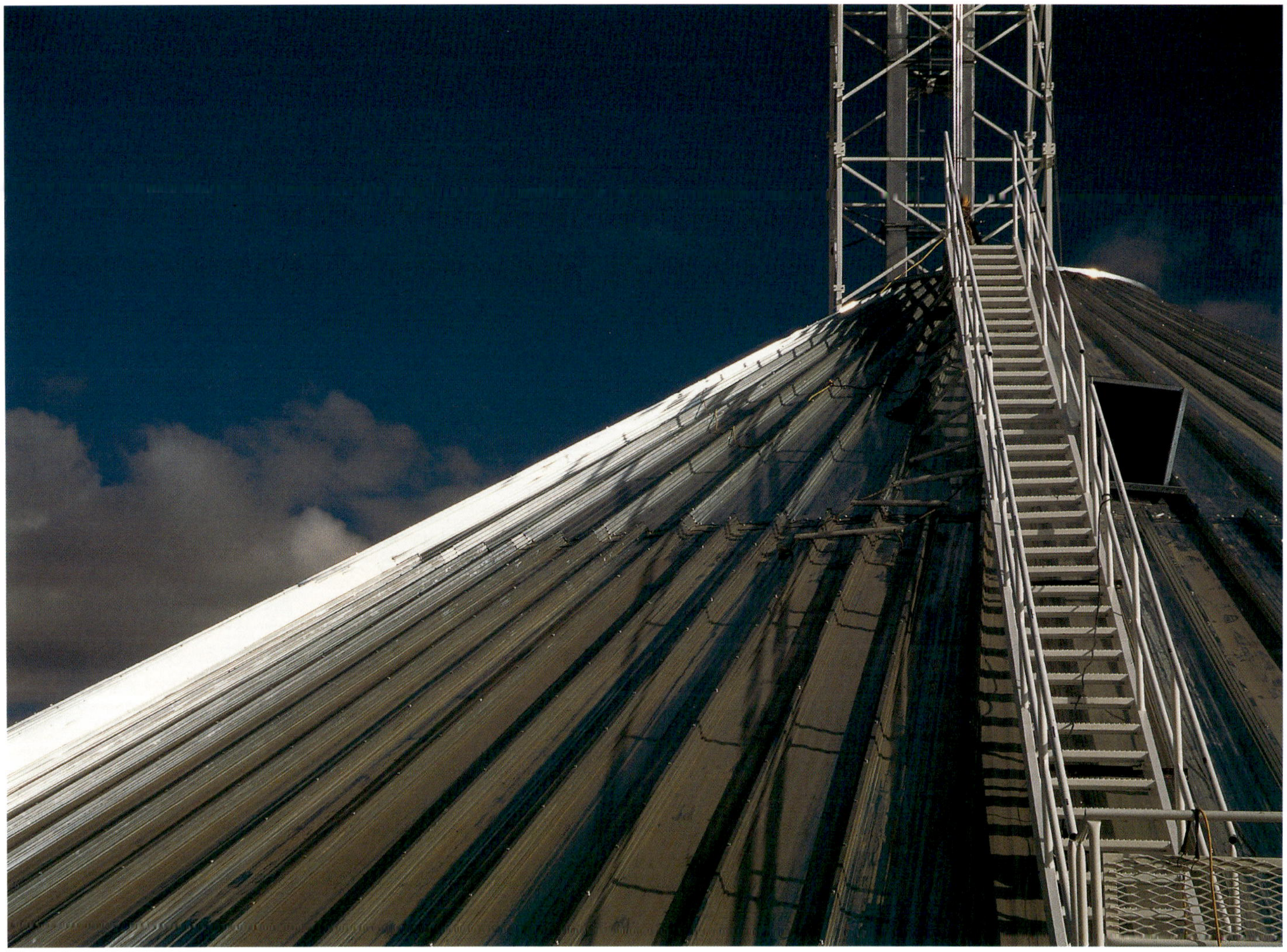

NEW ALUMINUM GRAIN BIN
Smithfield, Gosper County, Nebraska

the aquifer, a common pool resource, is being pumped dry at a considerable rate. Aristotle said, "What is common to the greatest number has the least care bestowed upon it. Everyone thinks chiefly of his own, hardly at all of the common interest." This phenomenon, the seemingly inevitable overuse of common pool resources, is called "the tragedy of the commons." But it is not inevitable.

Common pool resources are managed with conservation in mind everywhere on earth, from lobster fisheries in Maine to mountain grazing in Switzerland. But people using them have to take the first step, and acknowledge that their resources are limited. Not many farmers pumping irrigation water from the Ogallala Aquifer seem willing to take that step just yet. The wheat harvest was getting under way when I was in western Kansas. There were record crops (except for streaks of destruction where hailstorms had passed). What I saw was golden fields, combines, trucks, and not many people.

There are a few thriving market centers and many empty towns — falling-down houses where a family had once gathered for meals, some woman's precious lilac overgrown and dying beside the front porch — and always, as a backdrop, the towering concrete grain elevators alongside the railroad tracks. A few souls, often aged, go on in what are really ghost towns. Antique farm equipment rusts slowly into the ground, the sacred tools of a lost time. People make a hobby of visiting ghost towns in Kansas, examining the detritus of defeat. Signs on abandoned storefronts instruct,

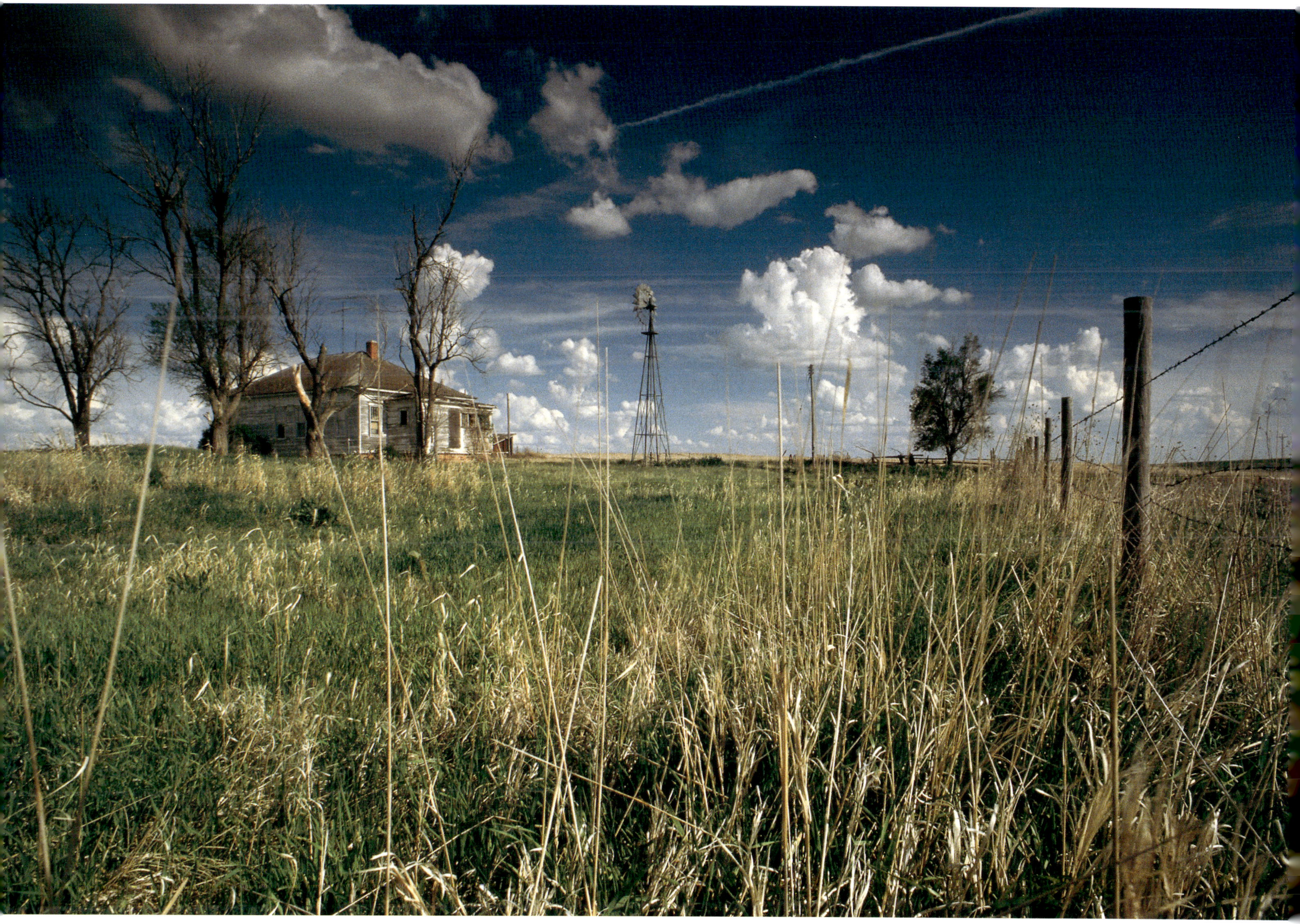

HOMESTEAD
North of Hendley, Furnas County, Nebraska

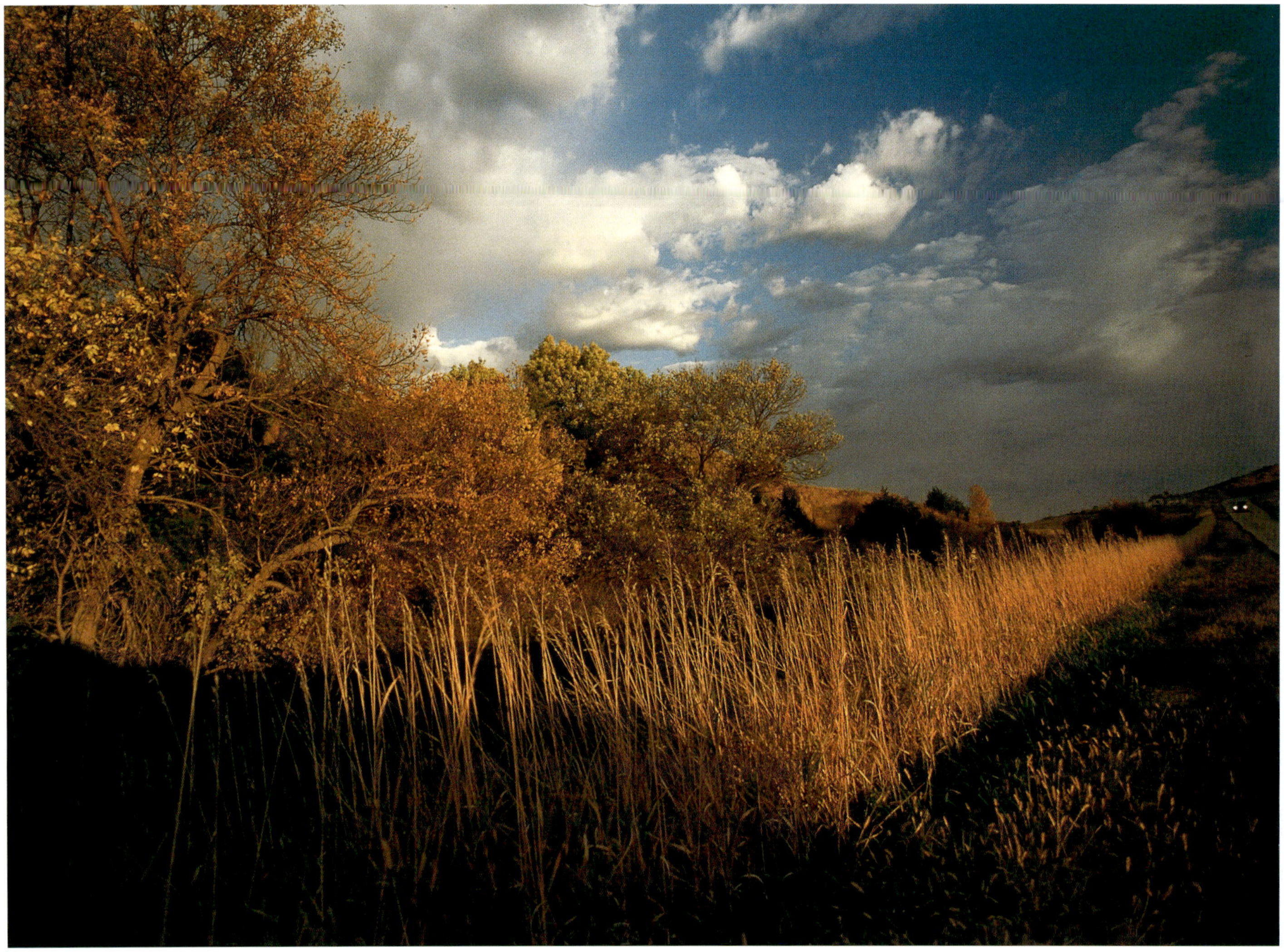

LONE PICKUP TRUCK
Custer County, Nebraska

"Last one out of town turn off the lights." In the tiny southern Kansas town of Bloom, I sought the friend of a friend, a man who raises goats, letting them feed on what once were lawns. He was not at home. Nor was anyone else, anywhere in town, so far as I could tell.

Back on Highway 83 I passed a bareheaded man walking south, carrying a stick and gesturing at the sky, miles from anything but fields of wheat. I thought of my father, gone to live overlooking the Pacific in old age, and his hours on his porch as he studied the weather and seas, the run of energies. I thought of going to one of these little semiabandoned towns, renting a house and waiting on the porch, getting used to the feel of infinities.

■

South of the Cimarron River, into Oklahoma, I was in drier, poorer country. Wheatfields vanished into sagebrush. By noontime it was 100 degrees; in coming weeks the drought and heat would never relent. Farmers in their fields, studying crops that simply shriveled on the stalk, got their photographs in *USA Today.*

Along the border between the Oklahoma and Texas panhandles, when the local culture looked to be most exhausted and numbingly dimensionless, I began to see welded sculpture made from tractor parts and the bumpers off old pickup trucks. Near Laverne, Oklahoma, there's a seventeen-foot replica of the Tin Woodsman. I wondered if it was intended as a statement about the uses of the imagination. What if the terrifying weather took us away, far from the monotonies of these sunbaked plains, to an Emerald City where adventurous lives were possible, and things happened in Technicolor. A girl might say, to her dog, some-

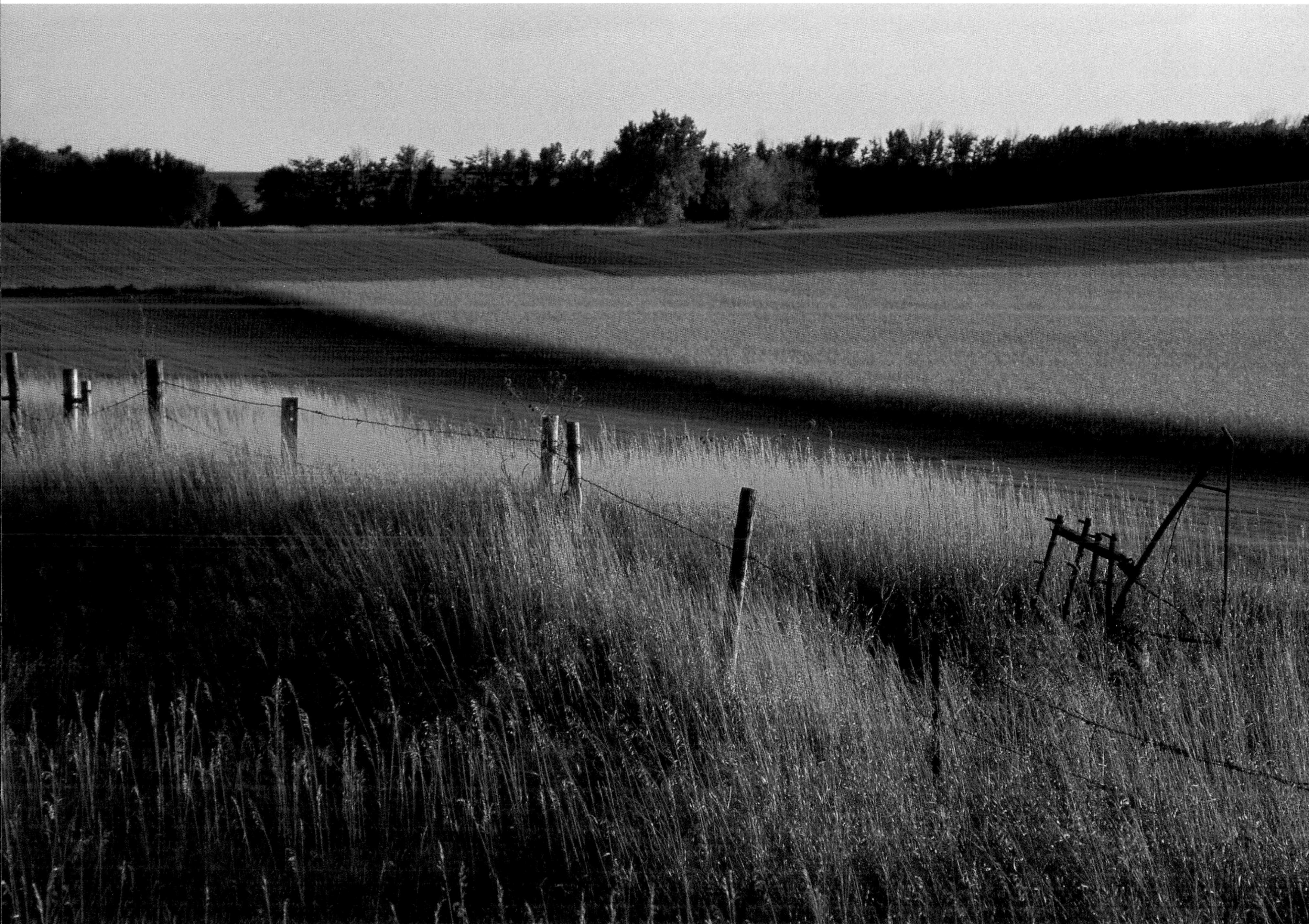

WHEATSCAPE

Near Stapleton, Logan County, Nebraska

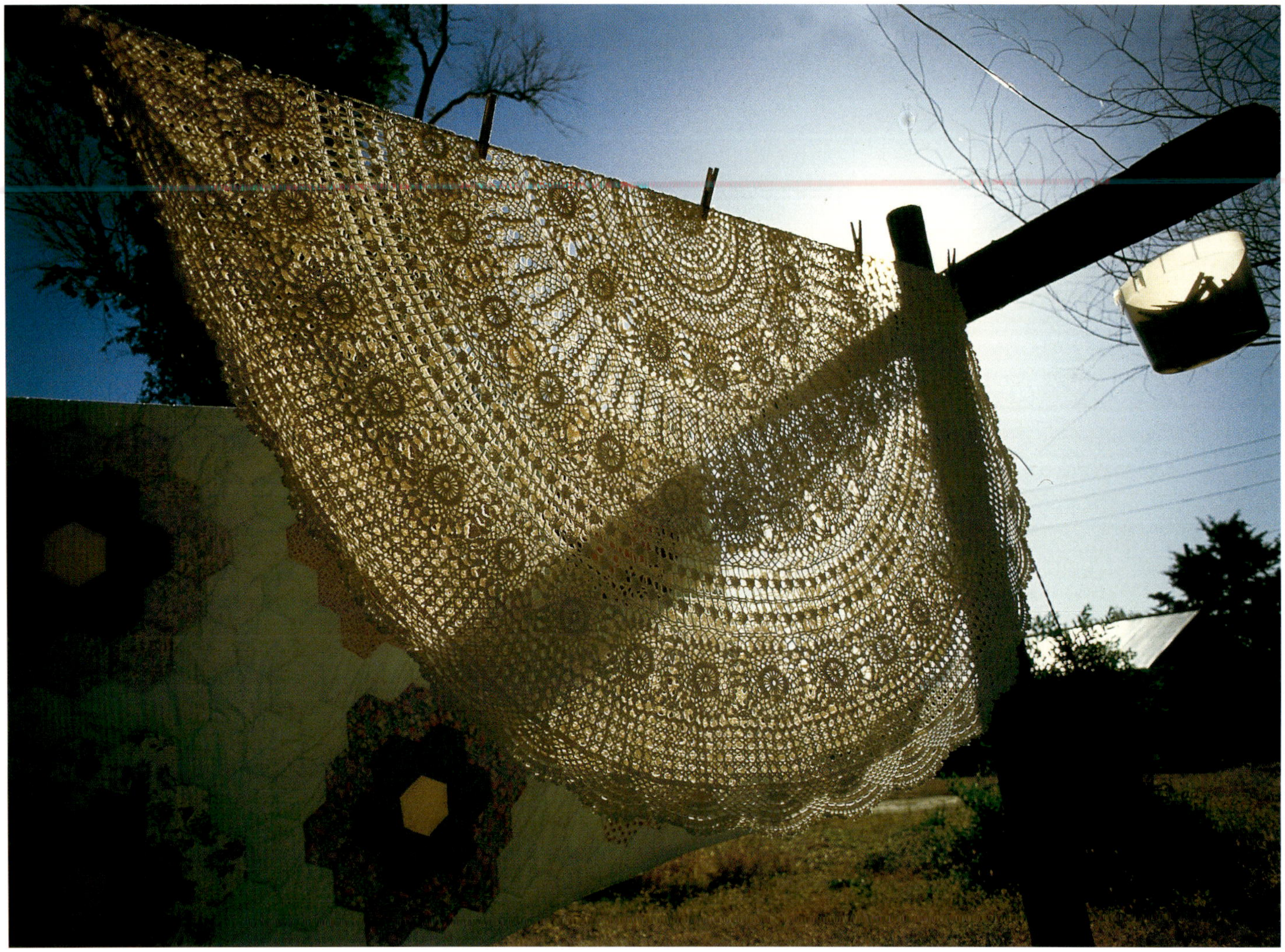

TABLECLOTH
North Custer County,
Nebraska

thing I recall as "Don't look like Kansas to me, Toto." She might be happy about it.

Jim Powers, a sculptor in Gage, has created figures, like an American Eagle made of car bumpers, that are strikingly ironic and sophisticated, as compelling as those in *The Wizard of Oz*. Artists do reinvent possibility. Jim Powers ought to be a public treasure in western Oklahoma.

The Black Kettle National Grassland looks much as it must have before everywhere nearby was cleared for agriculture. Tall grasses rolled in the wind and brush and trees were vivid green in contrast to the scars in the red earth that could once have been buffalo wallows. Was that grassland a remnant of the paradise that once reached everywhere on these plains? If so, we have lost more than we know.

The Washita battleground site, baking in the heat, is on the eastern edge of the grasslands. What took place at Washita is an American disgrace. In November 1868 George Armstrong Custer and 500 men of the 7th Cavalry massacred a defenseless Cheyenne village — camped exactly where they were supposed to be — while the regimental band played "Gary Owen." Custer, ever the fool, got everything wrong.

Another fellow on a motorcycle, a thick-handed man wearing bib overalls, came riding in. "I put a thousand miles on that bike this week," he said. I asked if he'd toured the Black Kettle grasslands. "Makes you see what all the trouble was about," he said. "Don't it."

■

Erick, Oklahoma, right on the border of the Texas panhandle, was the hometown of

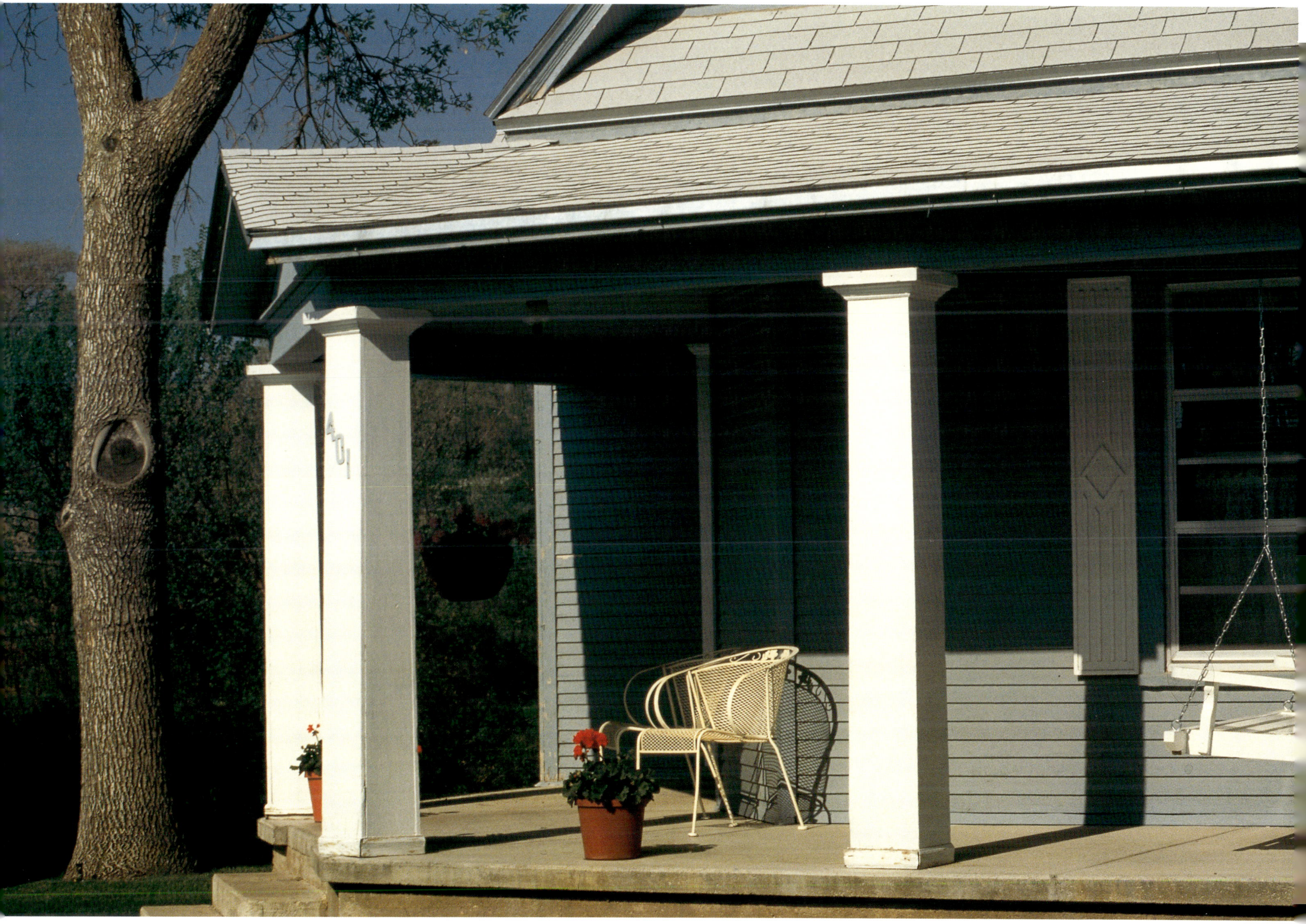

THE MOORE'S PLACE
Eustis, Frontier County, Nebraska

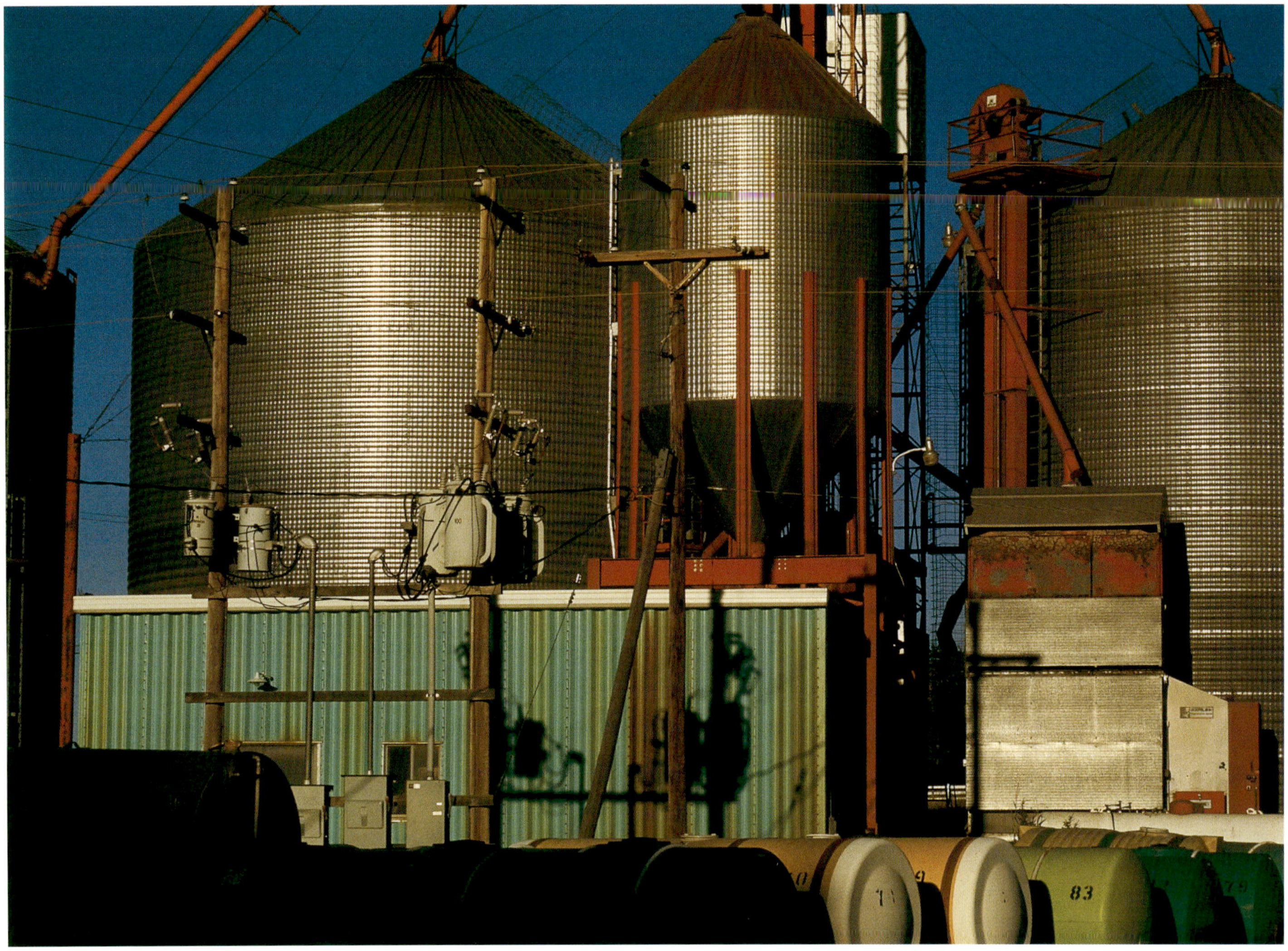

TANKS AND TRANSFORMERS
Merna, Custer County, Nebraska

Roger Miller, the country singer. "Trailer for sale or let, ain't got no cigarettes," was a poor boy lament we heard over and over like a Zen koan in smoky blue-collar taverns when I was younger, and taking my ration of heartbreak seriously.

Today, replaced by Interstate 40, U.S. Route 66 is a rough and deteriorating strip of two-lane concrete (it no longer exists as a numbered highway). In the 1930s it was a main road for the migration of dispossessed farmers and tenant families heading for California in their old Model T Fords during the Great Depression. What's left of it runs through the center of Erick, where it turns four-lane, with a median strip, and is called Roger Miller Boulevard when it intersects with Sheb Wooley Avenue.

Enormous storms were howling out of the Dust Bowl as folk in *The Grapes of Wrath* drove through Erick, heading for another promised land in California. When I was a boy growing up in the West, we called such migrants Oakies and Arkies; when the field work ran out, they went on relief; in the West they were treated with enormous disdain.

There's not much going on in Erick except for a 100th Meridian Museum. Not much of anybody on the streets. Maybe the great getaway took place long ago, or maybe everybody does their business over in Amarillo these days. Either way, it was empty windows in empty shops on empty streets in Erick.

After the Sunday desolations of little central Texas towns like Old Glory (a German-American community that took the name in order to seem patriotic during World War I — their high school is boarded

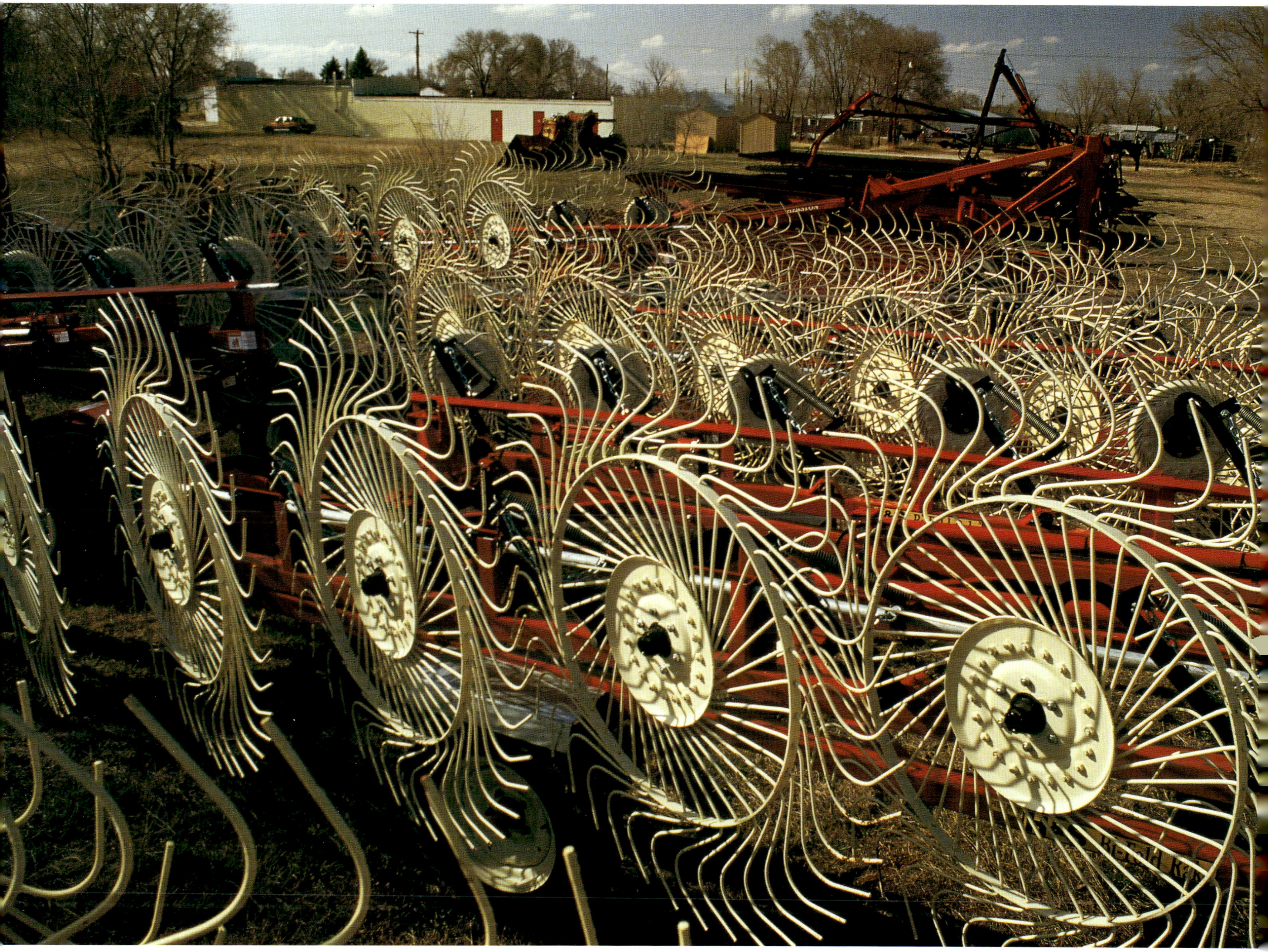

MACHINERY LANDSCAPE
Valentine, Cherry County, Nebraska

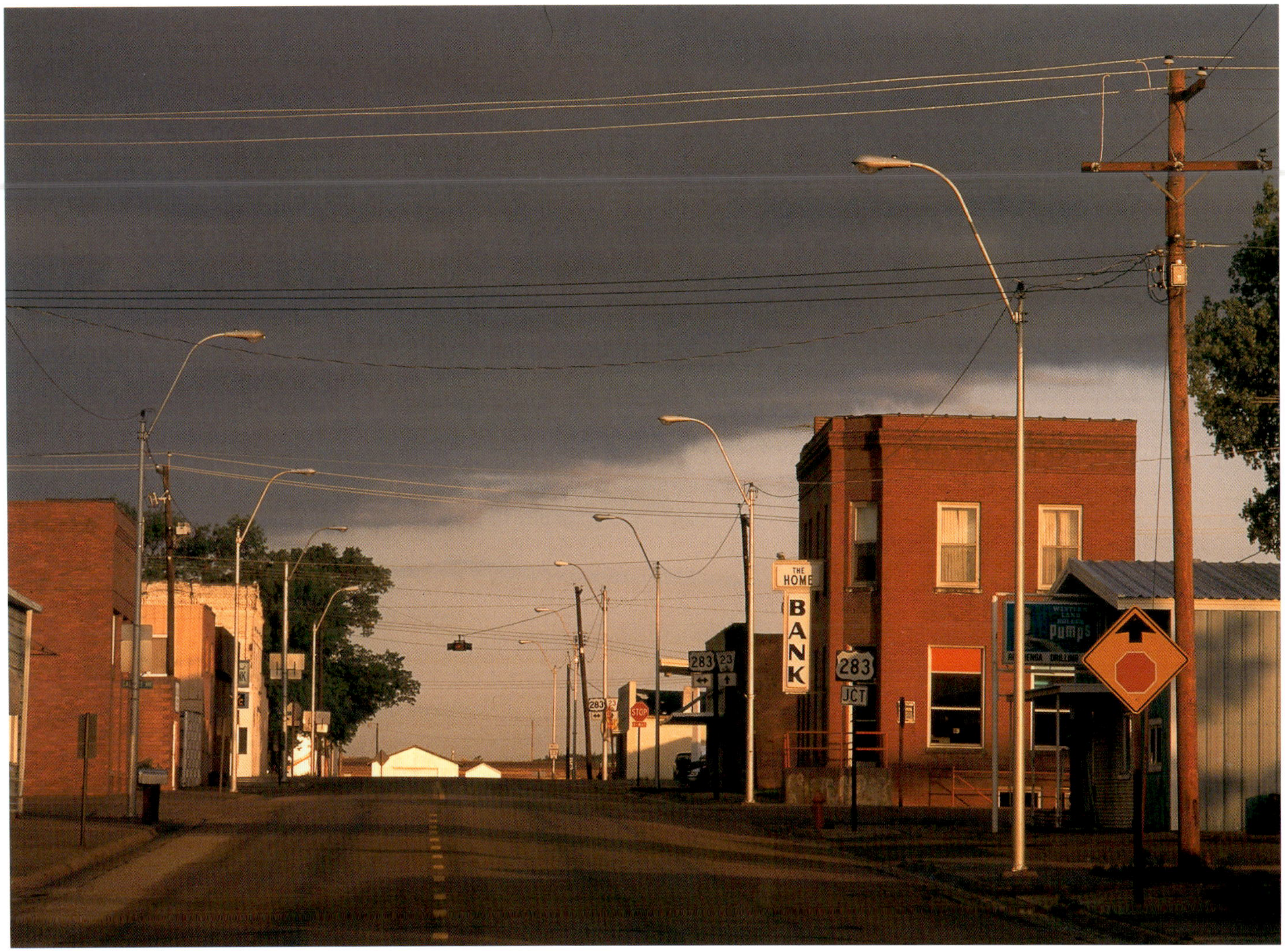

THE HOME BANK
Elwood, Gosper County, Nebraska

up), aimlessly driving the empty streets of Stamford, I was rewarded by finding museum-quality sculptures of dinosaur skeletons welded together from automotive parts by a body shop operator named Johnny Anders. Again, the transformation of reality by dream.

South of Stamford, crossing the Clear Fork of the Brazos, I was reminded of a moment years ago, at the Cowboy Hall of Fame in Oklahoma City, when eighty-nine-year-old Watt Mathews, with his older sisters, stood up to receive a lifetime achievement award. These were beyond doubt wealthy and powerful people; they owned a huge ranch on the Clear Fork; probably there was petroleum under their properties; they had never allowed anybody to drill for it; they lived a settled and ordered, feudal existence. Watt Mathews was a blue-eyed, solid little man who seemed to have been tempered to the hardness of steel. I wondered if he was like ruling class Westerners I've known, primarily dedicated to the preservation of privilege, the rewards his family had earned by enduring enormous difficulties, or if Watt Mathews and his sisters were among the rare ones, who understood that perseverance was only one step on the route toward inventing a sweeter world to inhabit. There are beautiful commercial buildings in once-upon-a-time cotton-empire towns like Ballinger — like the old stone feed mill, evidence that the founding citizens must have believed in grace as well as determination. I wondered what the people of Ballinger thought of just giving things away as a principle of life, a rule of civilized behavior.

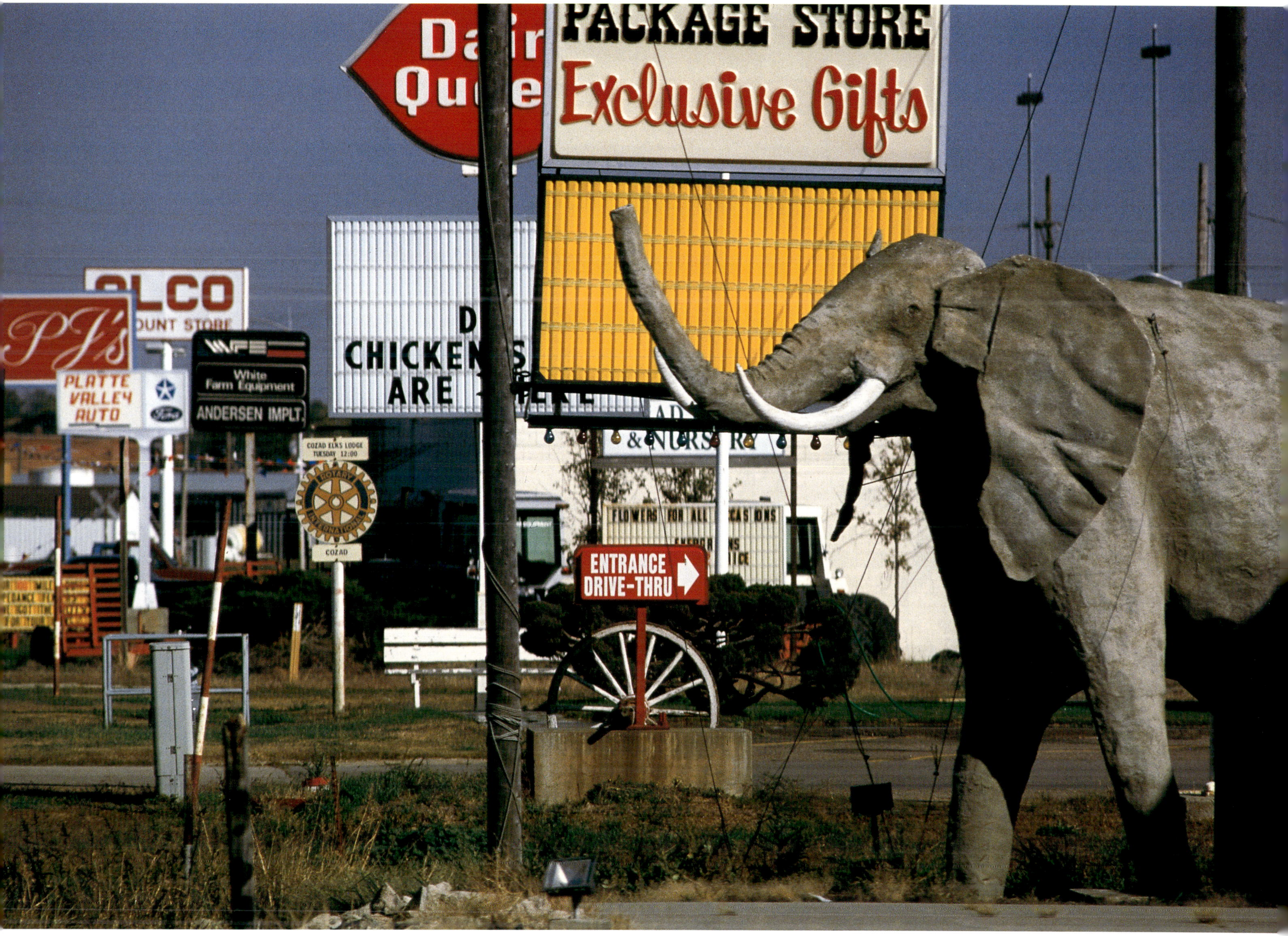

COZAD'S ELEPHANT
Cozad, Dawson County, Nebraska

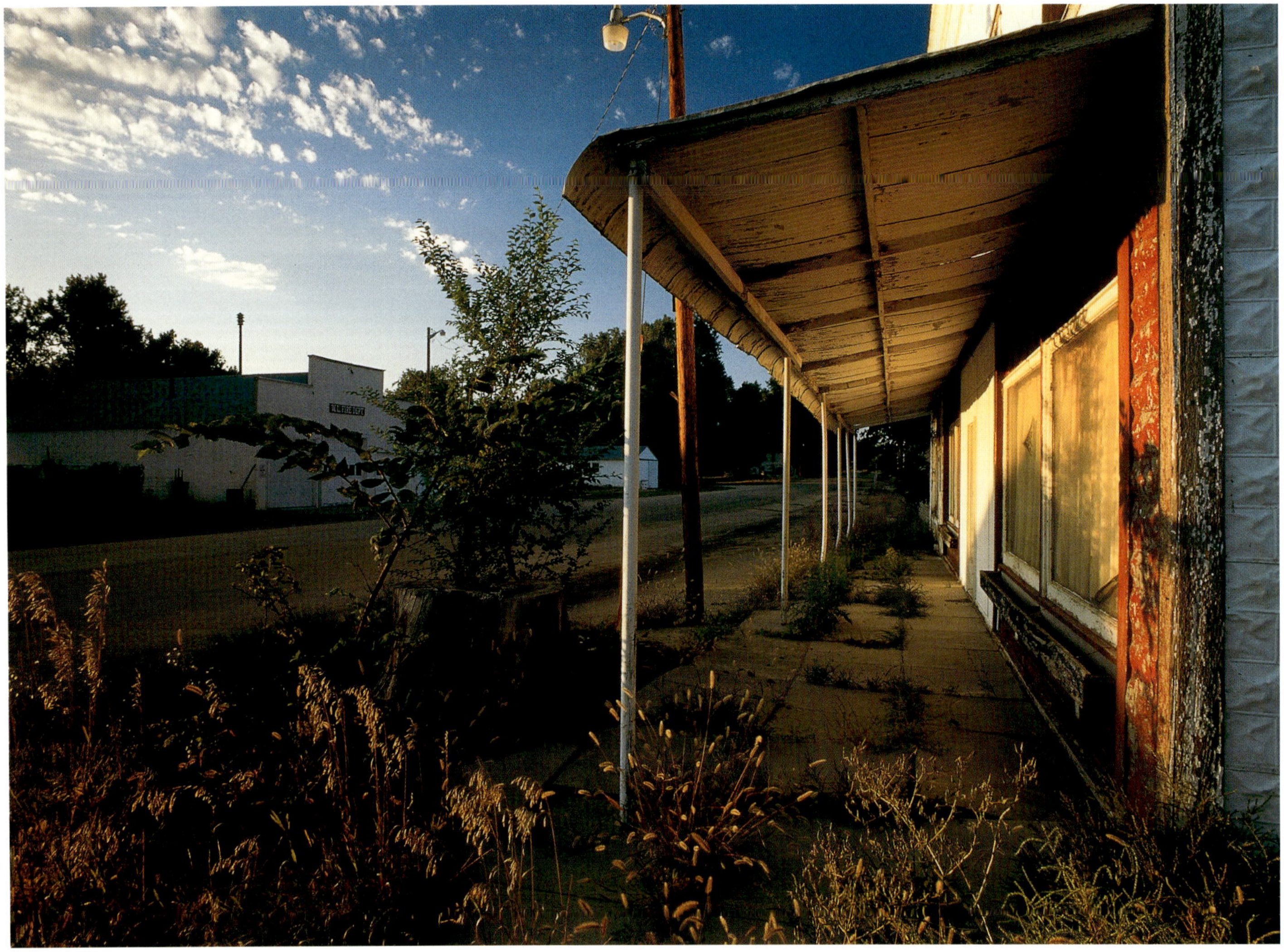

STREET SCENE
Wood Lake, Cherry County, Nebraska

When, I wondered, would I encounter the legendary southwestern lightness of the heart and wildness? I searched for barbecue and I longed for heedless songs, some transcending, and dancing around, maybe balloons on the ceiling, even a martini.

The Poetry of Force

The history of life on the southern Plains, so much as we know of it, the part that can be called our history, began with hunting and gathering peoples and Spanish horses.

The Comanche were obsessed with horses; they were the finest horsemen and most skilled Plains warriors. The story of their rise and fall resonates with what Simone Weil, talking about the heroes in the *Iliad,* called "the poetry of force." The Comanche drove the Apaches out of Texas, raided the whites and their black slaves, expanding into Texas. Deep into the Republic of Mexico, they killed and looted; they drove herds of stolen horses up from Mexico on the old Comanche Trace, which crosses the Rio Grande just west of the 100th meridian; war sustained their empire.

On May 19, 1836, Comanche attacked the little settlement known as Fort Parker, on the cutting edge of white advance into east-central Texas, and kidnapped John Parker, aged six, and Cynthia Ann Parker, aged nine. John Parker was ransomed back from the Comanche in 1842, when he was twelve. Soon he ran off to the plains again, seeking and never finding his sister, at last settling in the desert country of northern Mexico. Cynthia Ann became a Comanche wife and bore two children. On December 17, 1860, she was recaptured; she had for-

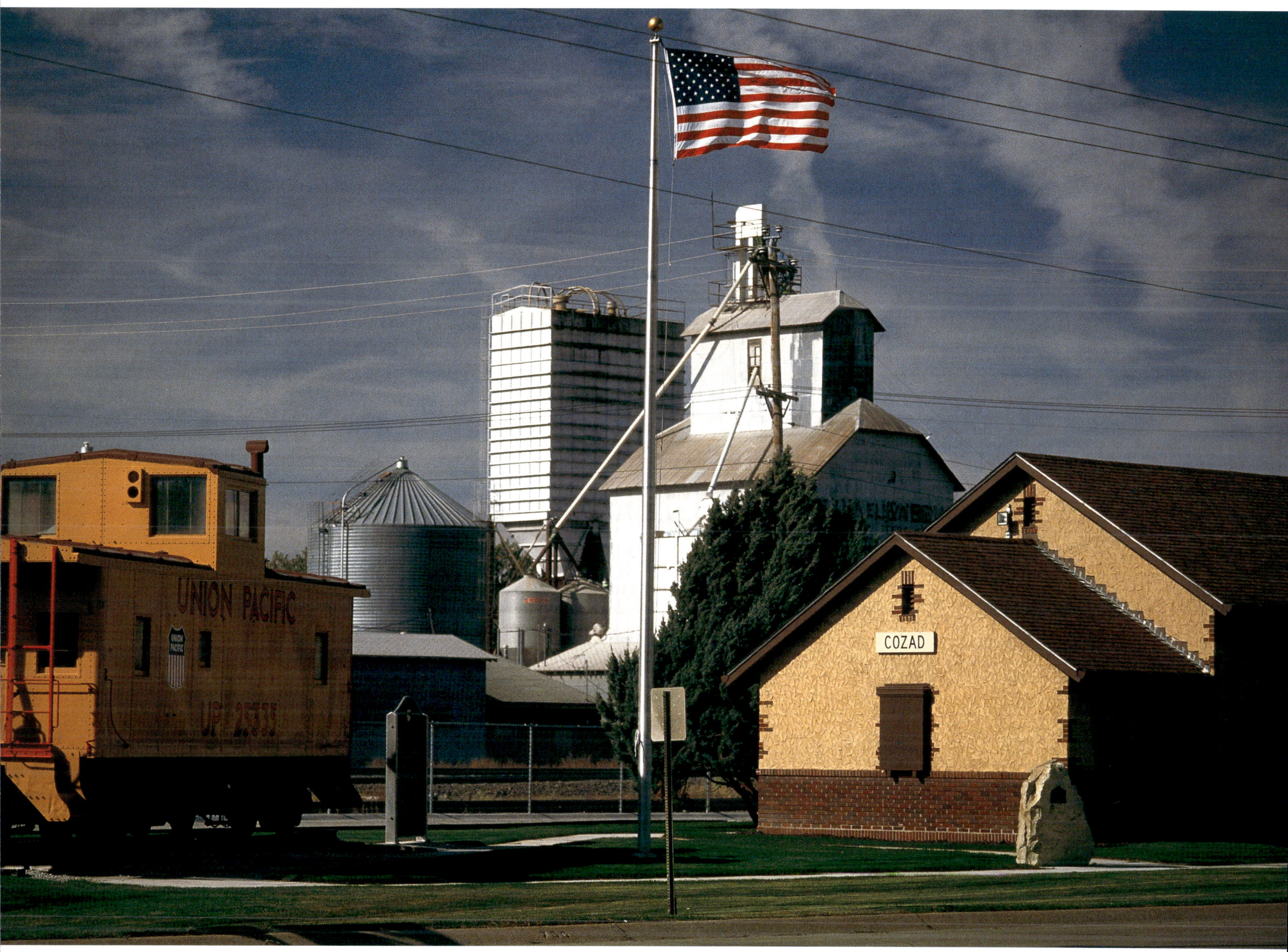

COZAD STATION
Cozad, Dawson County, Nebraska

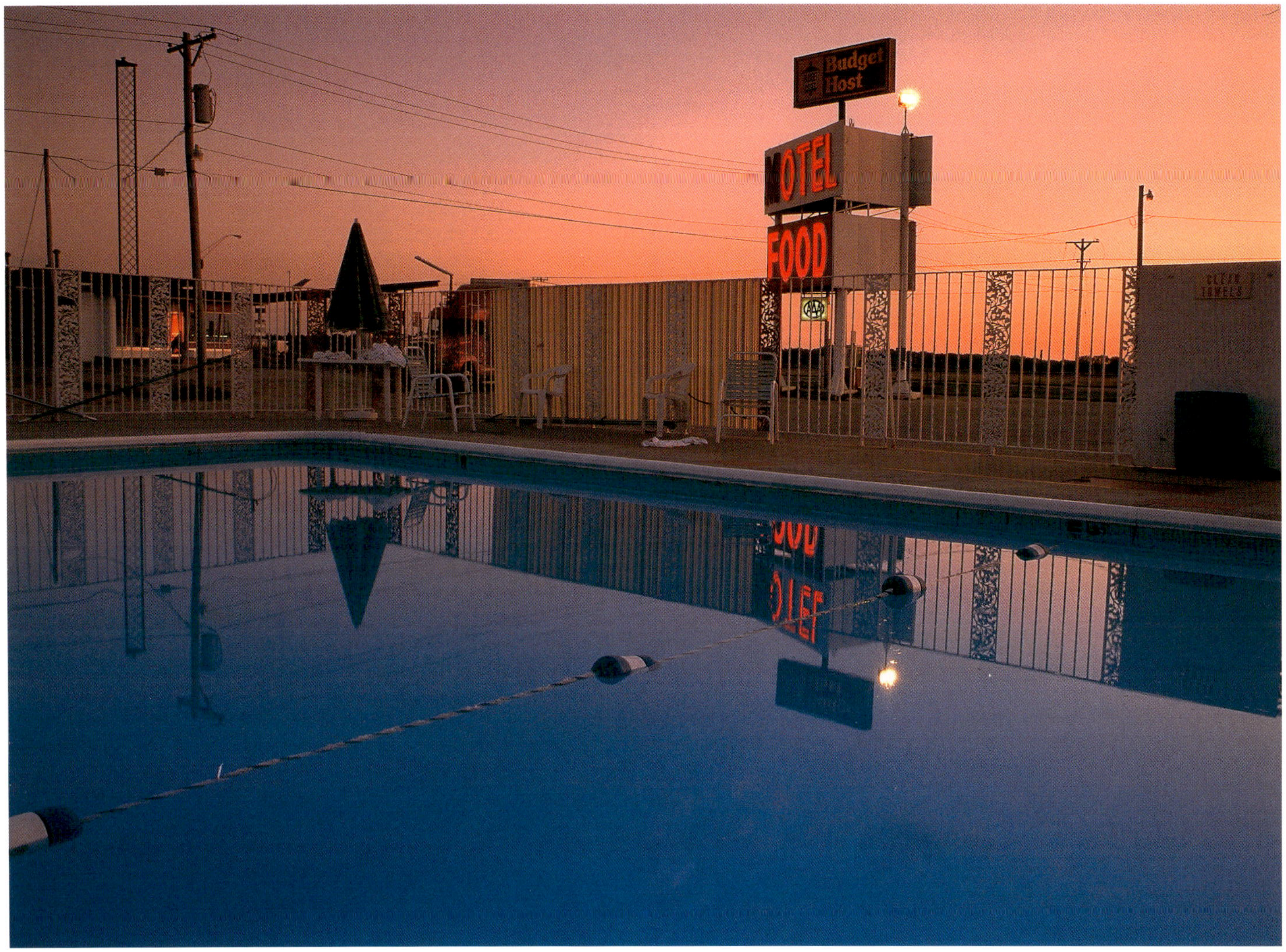

HOT NIGHT
Dawson County,
Nebraska

gotten life in the white settlements; she was Naduah, she had been a Comanche woman for twenty-five years. Her husband and her sons were still out on the prairies (One was Quanah Parker, known as Sweet Odor, who retained his mother's name as a mark of his respect for her. He was the last of the Comanche war chiefs, a friend of Teddy Roosevelt who oversaw movement of the Comanche to reservation life). Naduah tried to escape and go back home to the Comanche; she was put under guard. Her young daughter died; it is said that she howled and mutilated herself, sank into apathy, and starved herself to death. It is, in any event, an utterly Homeric story.

■

When the Civil War ended in 1865, there were millions of long-horned cattle wandering the brushy flatlands of south Texas, mostly unbranded and free for the gathering. They could be turned into money in eastern markets. The Santa Fe Railroad was beginning its march across Kansas. The great trail drives began, pushed ever westward by encroaching settlers and the wire fences, to the legendary cowtowns like Abilene in the 1860s, and Dodge City by 1874 (in 1877, with a population of less than a thousand, there were sixteen saloons in Dodge City; by 1885 the trail drives to the railhead had been forced farther west, and Dodge City's years as Queen of the Cowtowns was over). The Santa Fe hauled farmers west into Kansas, hauled cattle from Texas, and enormous numbers of buffalo hides, and bones, to eastern markets.

In 1868 the southern herd of buffalo was still immense. Estimations of the num-

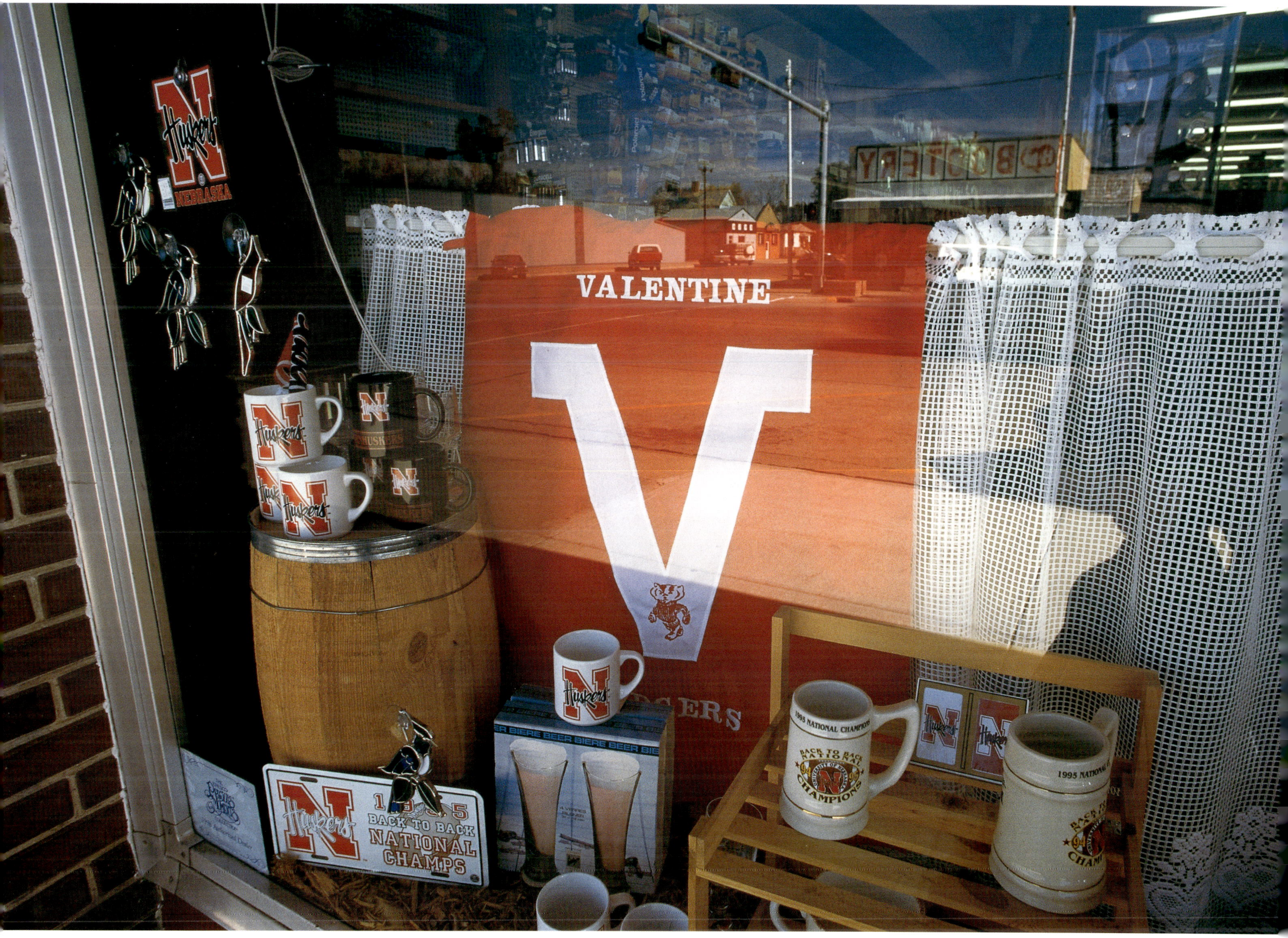

GO BIG RED!

Valentine, Cherry County, Nebraska

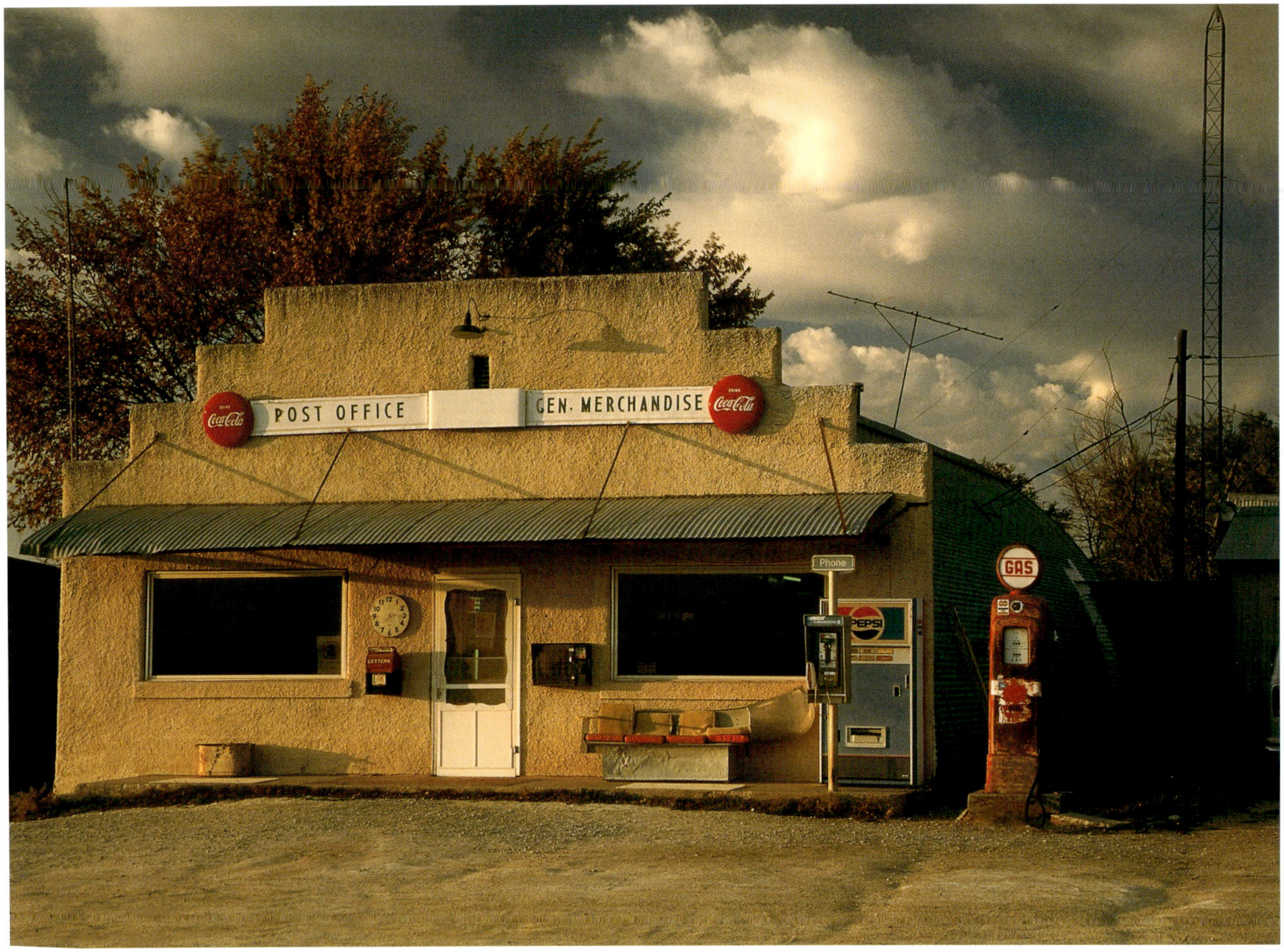

SPARKS STORE
Sparks, Cherry County, Nebraska

bers are guesswork, but it's certain there were more than 35 million animals. Hunters killed them in just a few years, taking mostly hides. Glistening carcasses rotted in the sun. During 1872–74, one and a half million hides were shipped from Dodge City alone. In the winter of 1877–78, only 100,000 buffalo were left to be killed. By 1881 the bones of 31 million had been sold for fertilizer, and after that no more records were kept. The bones and the buffalo were gone.

How to think of that conjunction of cultures — wheat-farming Mennonites from the fringes of Asia, cowboys out of Texas, blacks fleeing the South, the buffalo hunters, vast stacks of hides, mile-long piles of bones alongside railroad sidings, and eventually those six-story wheat elevators riding like tall ships on the sea of the plains? And the Indians, like the Comanche, who were witness to the utter transformation of their homeland while enduring final defeat and subjugation.

In 1875 Ranald Mackenzie, the most implacable leader among the Indian fighters in the U.S. Cavalry, trapped the last great Comanche war party, led by Quanah Parker, in the depths of Palo Duro Canyon, just south of Amarillo, and captured all their horses, 1400 head. Some of the horses were given to scouts as a reward; more than a thousand head were shot. A trader, years later, hauled their bones away to sell as fertilizer.

The Comanche, a culture of horseback warriors if nothing else, were immobilized, on foot, and thus absolutely defeated. They surrendered unconditionally. Quanah

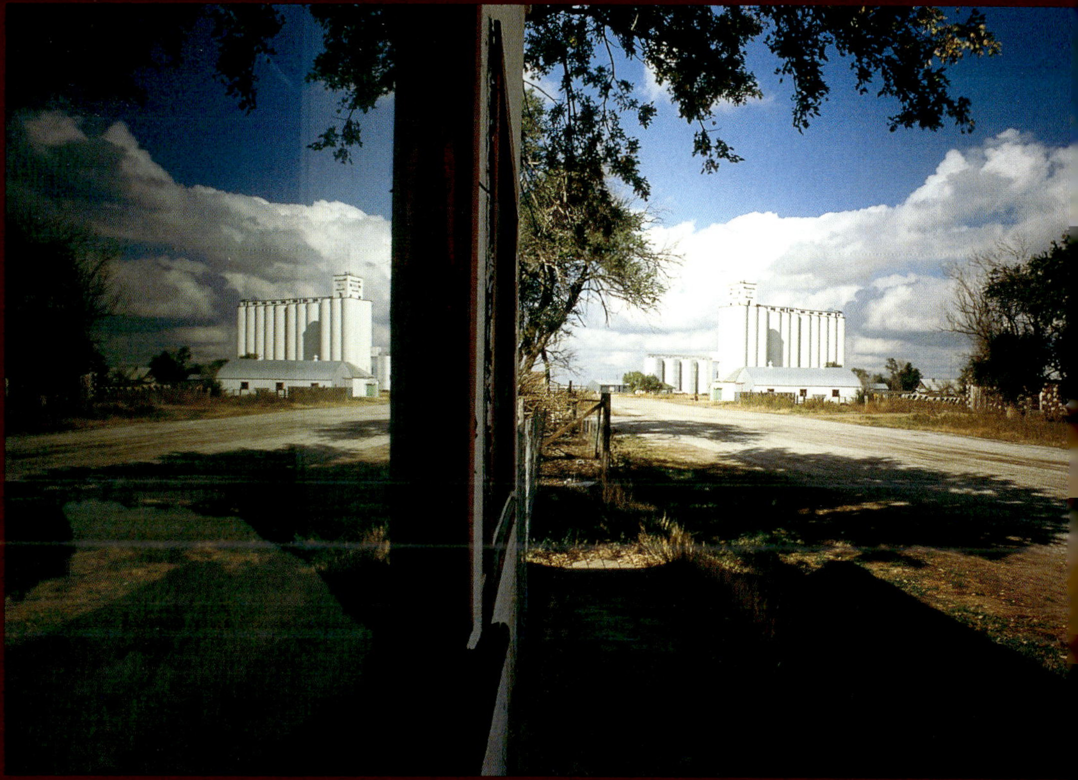

K A N S A S

AMERICAN REFLECTIONS

Bloom, Ford County, Kansas

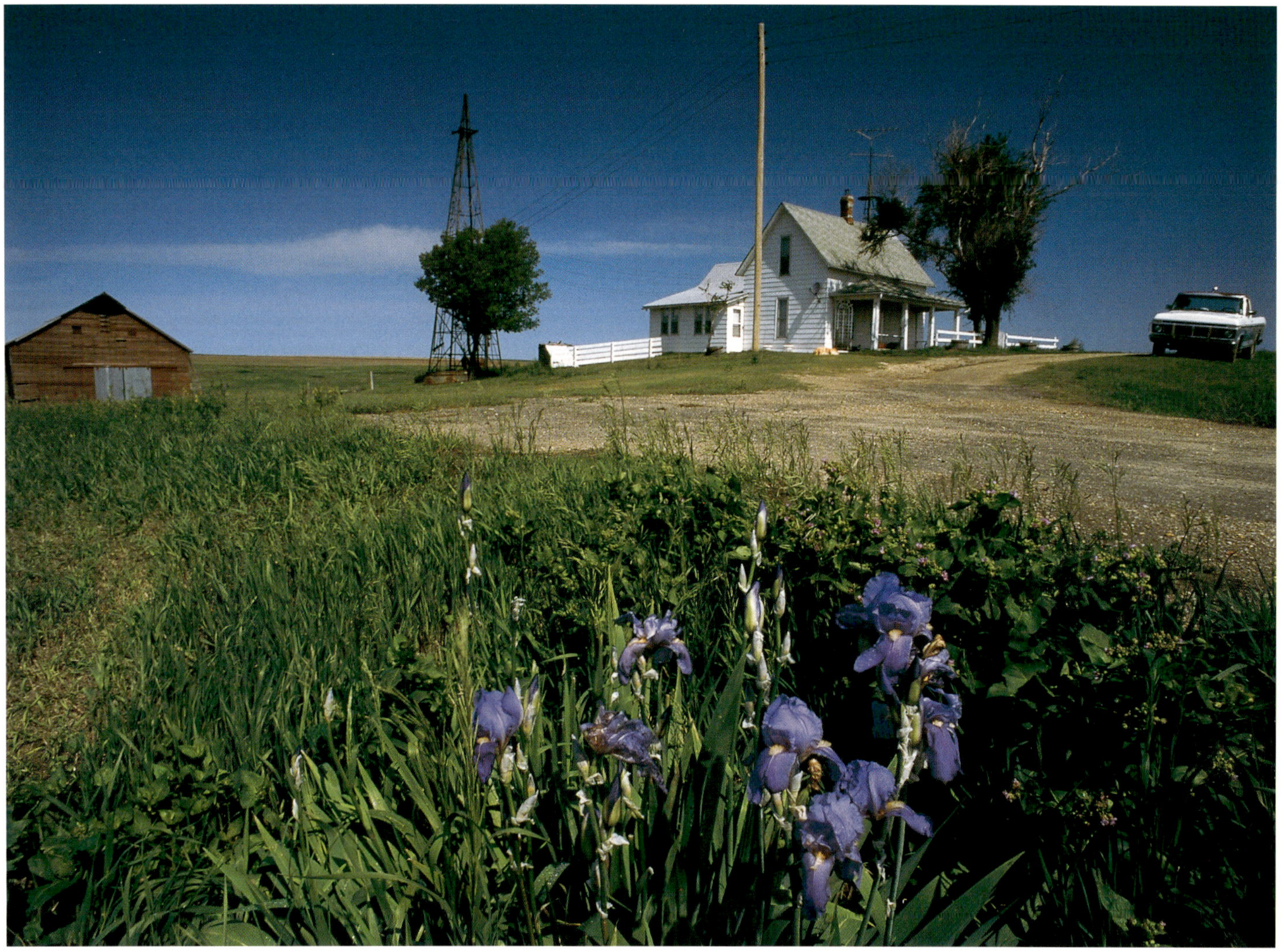

FARMHOUSE WITH IRISES
Norton, Norton County, Kansas

Parker led them to their reservation in Oklahoma. Three years later, in 1878, after Comanche leaders asked permission to hunt buffalo on the plains, fifteen hundred warriors rode out with a military escort. They found no buffalo; they killed horses for meat. There was nowhere to go but back to the reservation. The Jeffersonian agenda, settlement of a farmer's homeland, could now proceed across the southern plains. There was only the weather to worry about. And the infinities.

The Next Picture Show

Centers of innovation in human cultures have traditionally been in cities. But that didn't seem so down the 100th meridian — not in Bismarck or Pierre or McCook, Nebraska, or in Dodge City, Kansas, or Abilene in central Texas. Those small cities seemed devoted to the status quo.

The sacred work of our society has lately centered on relentless commodification. Laredo, at the U.S.–Mexican border on the Rio Grande, is a little city on the cultural cusp, where the ancient energies of family driven rural Mexican culture and the franchise culture that is deadening so many American environs. It was ringed by a run of franchise businesses like Tony Roma's Famous Ribs and Long John Silver's, Sam's Club, and Fuddrucker's. In the Texas Highway Hospitality Center, on the freeway leading into Laredo, I encountered a man with his possessions in plastic bags. Surrounded by malls, he muttered to his muse. Another day of drowning.

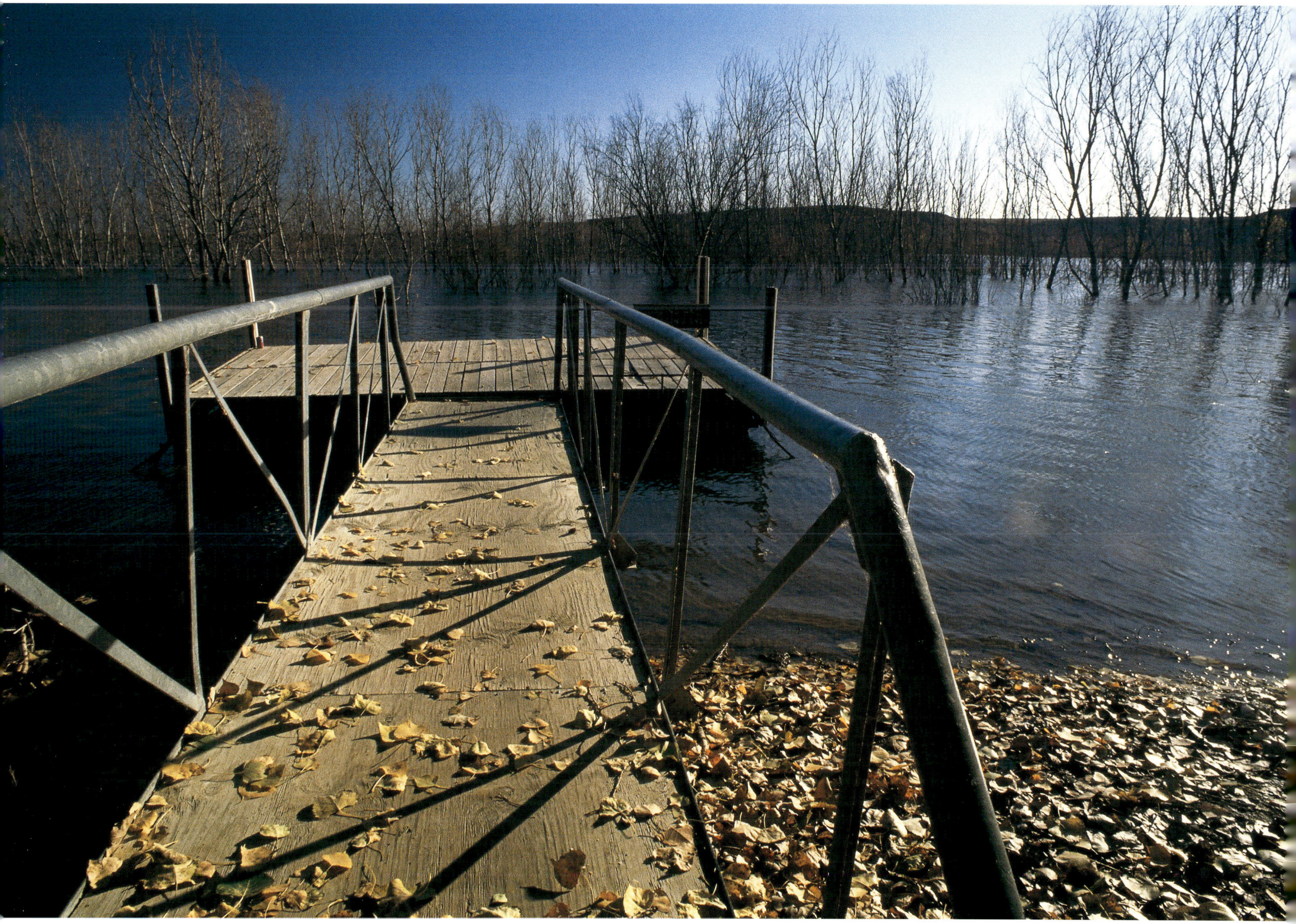

CEDAR BLUFFS STATE PARK
Trego County, Kansas

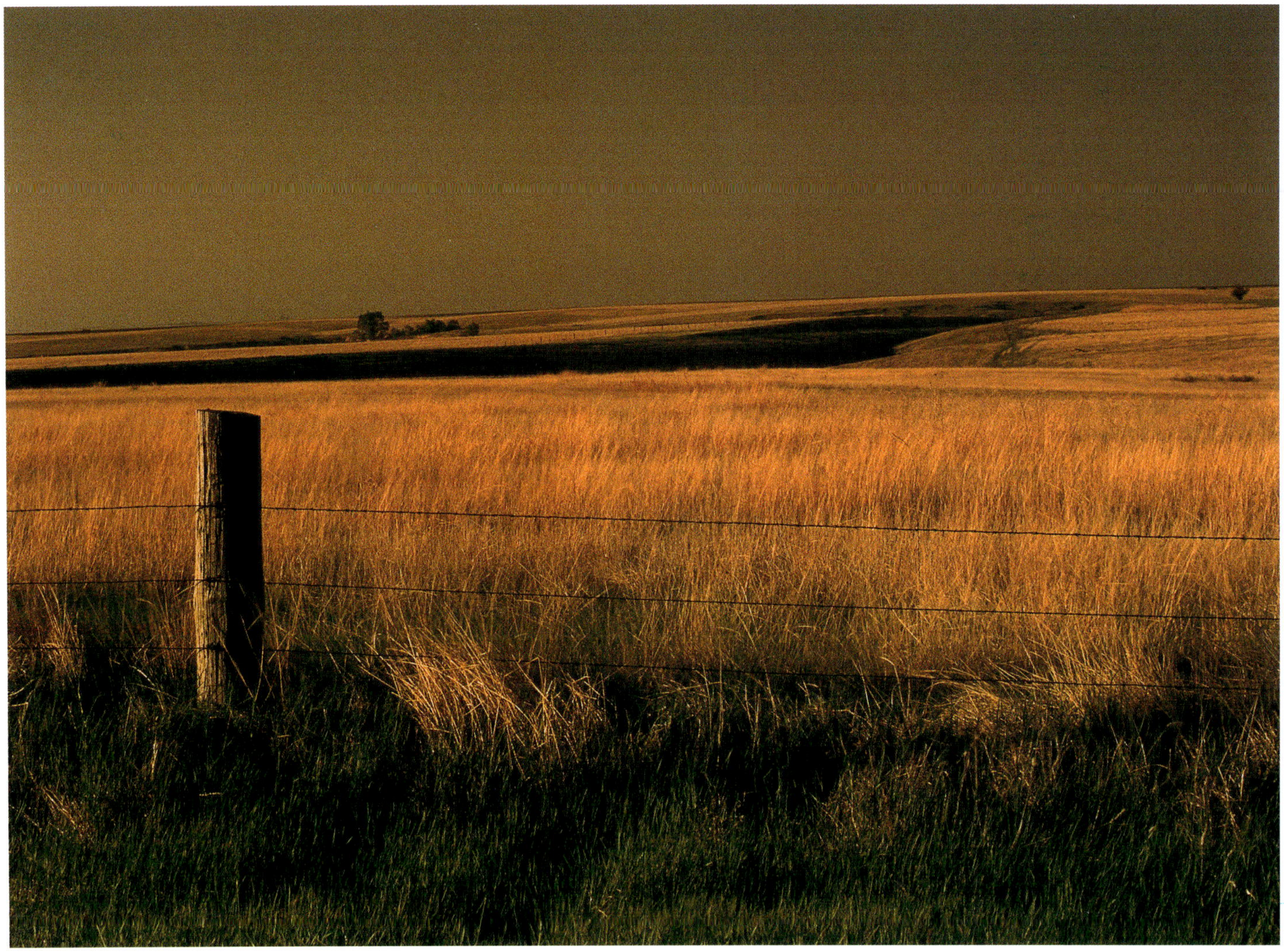

FENCE LINE
South of Jetmore,
Hodgeman County,
Kansas

Laredo is the largest inland port in the nation, transshipment point for 40 percent of the goods coming into the United States from Mexico, a city-sized warehouse full of junk pottery, paintings on velvet, rawhide furniture, reindeer made of woven wire, cast-iron patio furniture, plaster replicas of the Statue of Liberty, and suits of pseudo-medieval armor made of tin cans painted with varnish.

In San Augustin Square — the heart of the downtown — everybody was sweating together and it was 103 in the shade by eleven o'clock. San Augustin Square is primarily just another merchandise mart, but unlike the malls around the edges of Laredo it was vividly inhabited. It is a sort of ceremonial center, burning with cross-cultural energies. The pleasures of sociability and give-and-take between strangers seem to be continuously celebrated.

Mexican agricultural towns to the north, like Crystal City, are populated by people who are eager in their willingness to work in pursuit of the American scenario — economic survival and psychic freedom. Migrant Hispanic workers go north en masse from Crystal City to Minnesota each spring, often six or eight to an old beater car, and work their way south through the summer, picking crops.

That's how it was there on San Augustin Square, people sweating and hustling and smoking and loafing and laughing, doing the old human dance, a neighborhood society running on forbearance. They haven't erected a barricade made of commodities against one another. If Americans had any sense, the wheelings and dealings,

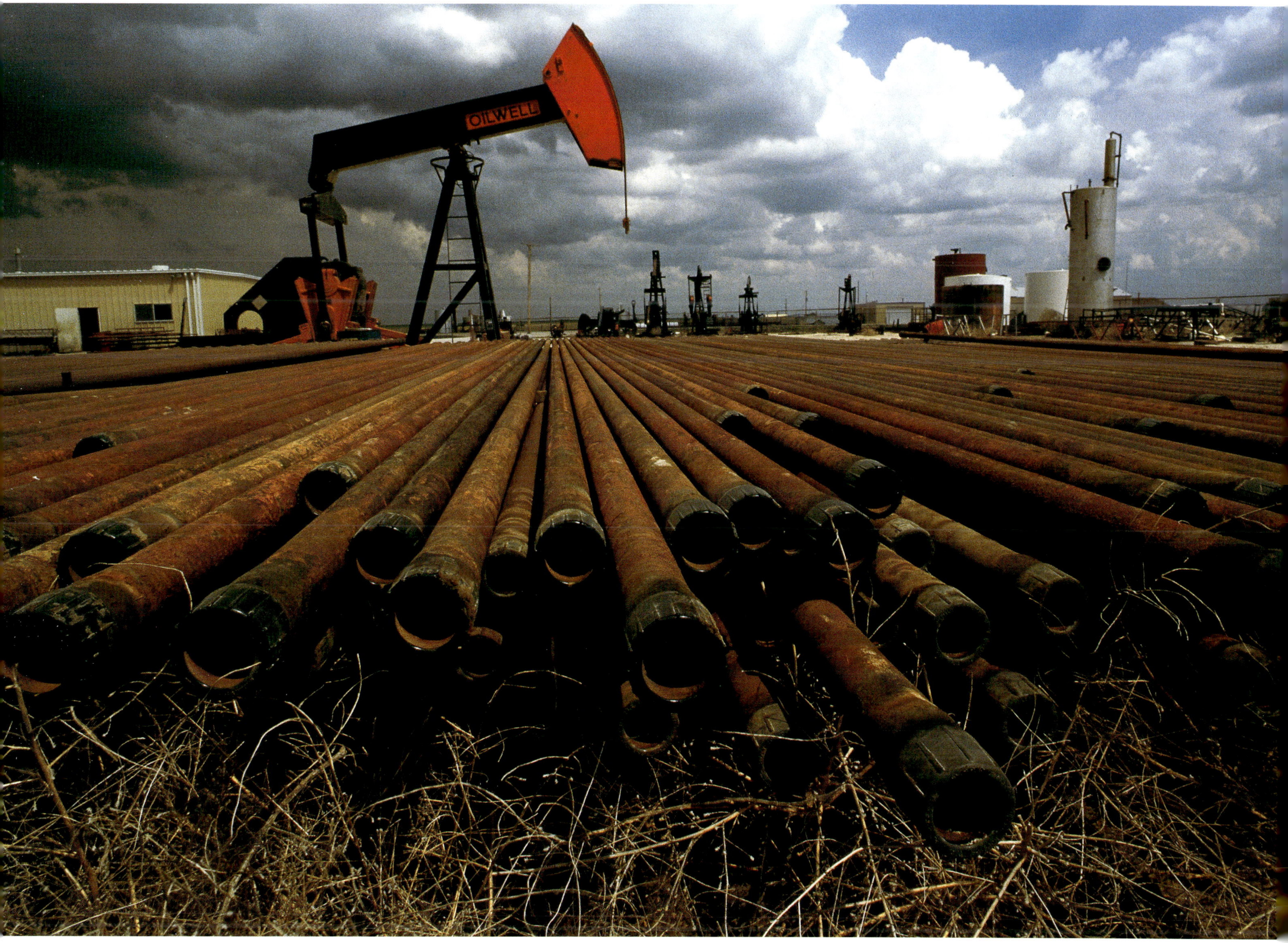

PETROLEUM LANDSCAPE
Ness City, Ness County, Kansas

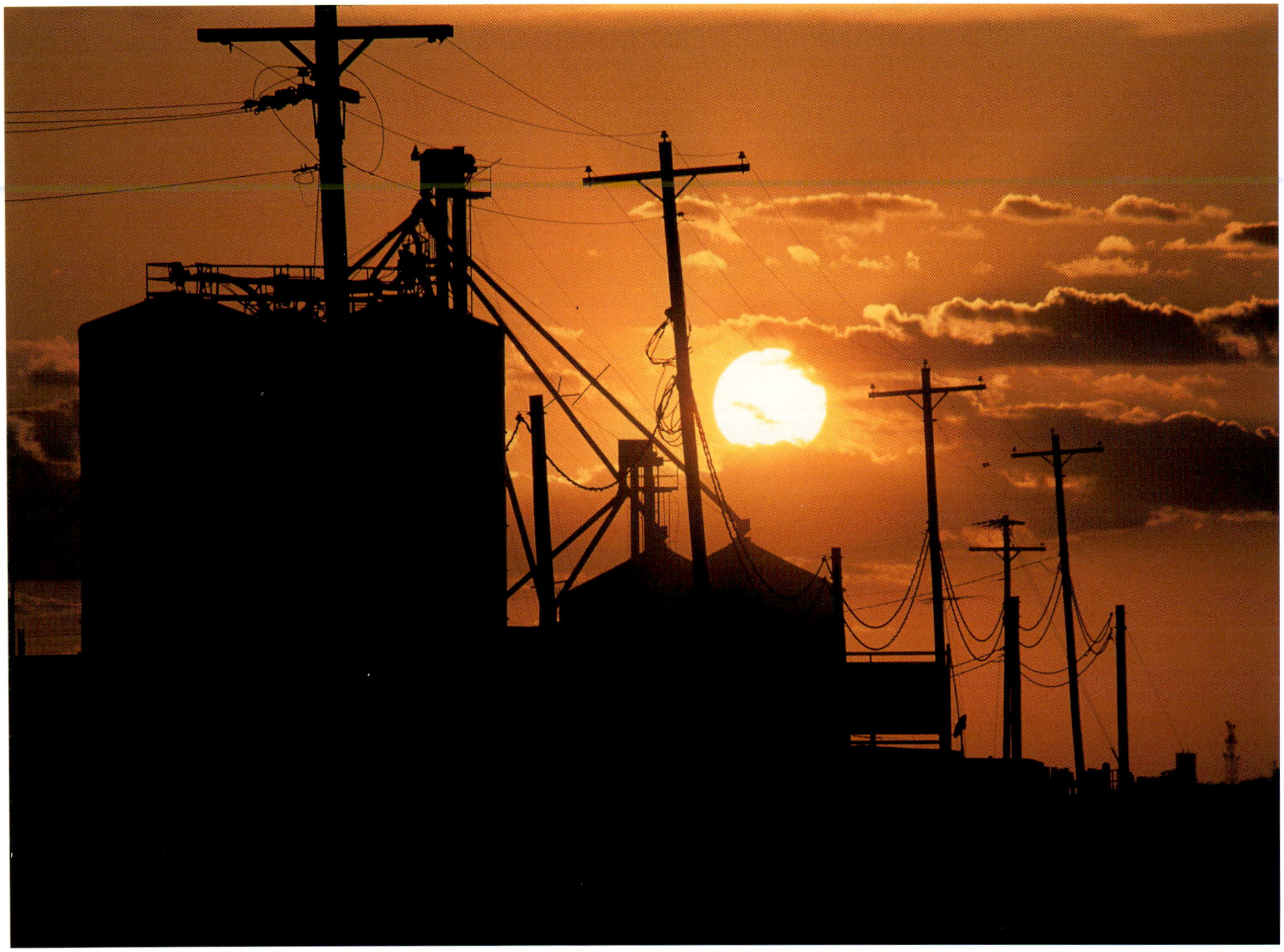

STORAGE BINS
Dodge City, Ford County, Kansas

the laughter and guitar playing, buying and selling and trading, the mixing on San Augustin Square might stand as a model for our multi-ethnic future. Throughout the world as it shrinks, our cultures increasingly have to live with one another. We are creatures with nowhere to be except on this planet, sharing the fertilities.

■

Maybe survival on the plains was too difficult. D. H. Lawrence wrote, "The savage America was conquered and subdued at the expense of instinctive and intuitive sympathy of the human soul. The fight was too brutal."

In the 1980s, preparing to write a film script, I had occasion to watch hours and hours of 8mm home movies shot over decades by a man who had lived his life in a North Dakota town settled by German people whose ancestors came to America after centuries of farming in Russia. I saw babies in arms, and the same babies when they went to school, and later in life with children of their own. I watched those people plow great drifts of snow after horrific storms, I watched their weddings, feasts, burials, I saw people admiring new automobiles, children leaving for college. I saw real people aging and learning to fend off what James Agee called "the enormous assaults of the universe."

I hoped to find, along the 100th meridian, a society of forthright giving citizens, an earned communal culture. And I did. But a wounding had occurred during the difficult years; it doesn't seem to heal. Settlers who lasted on the Great Plains understood that they were survivors in a

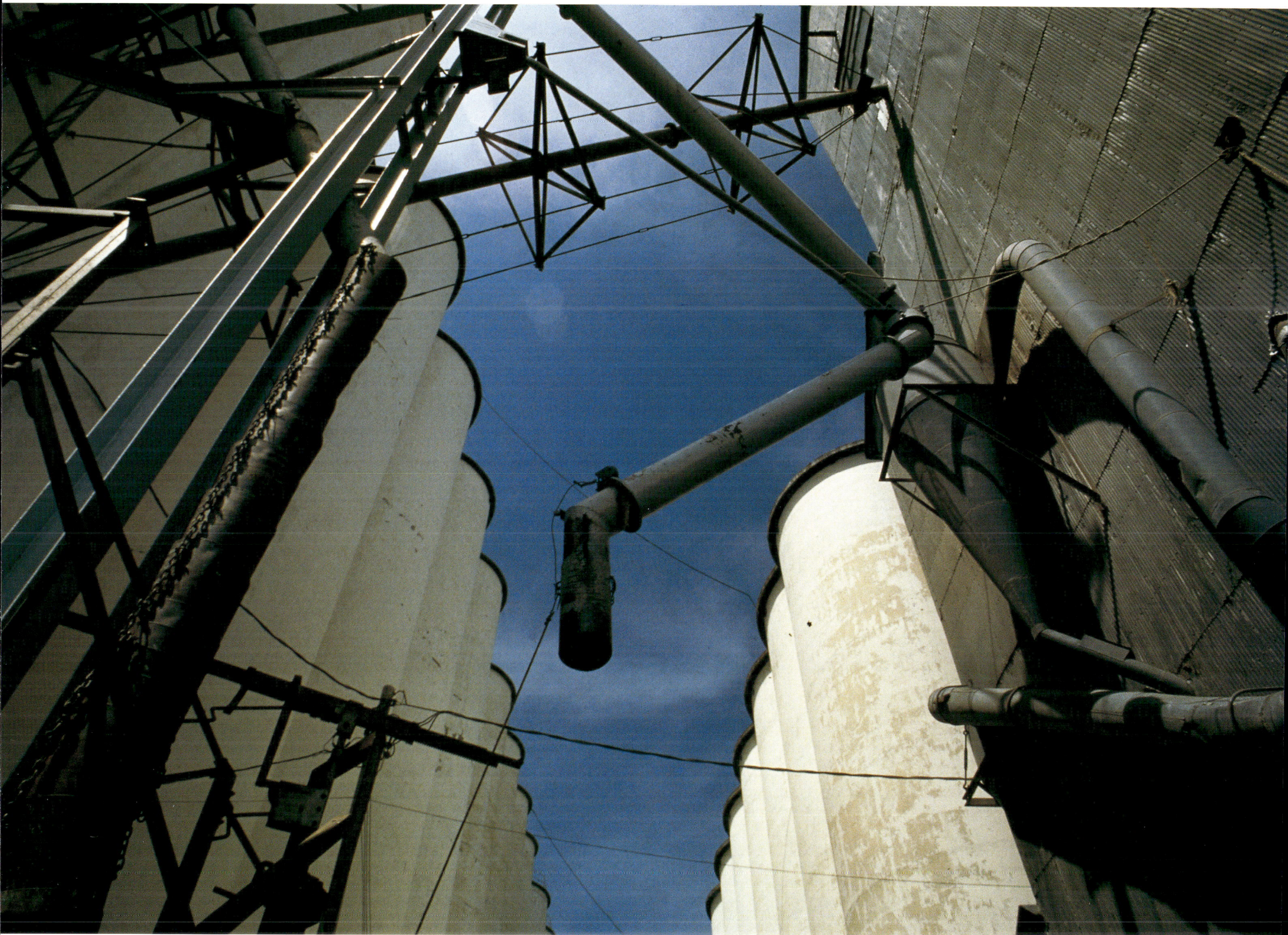

BONDURANT GRAIN COMPANY
Ness City, Ness County, Kansas

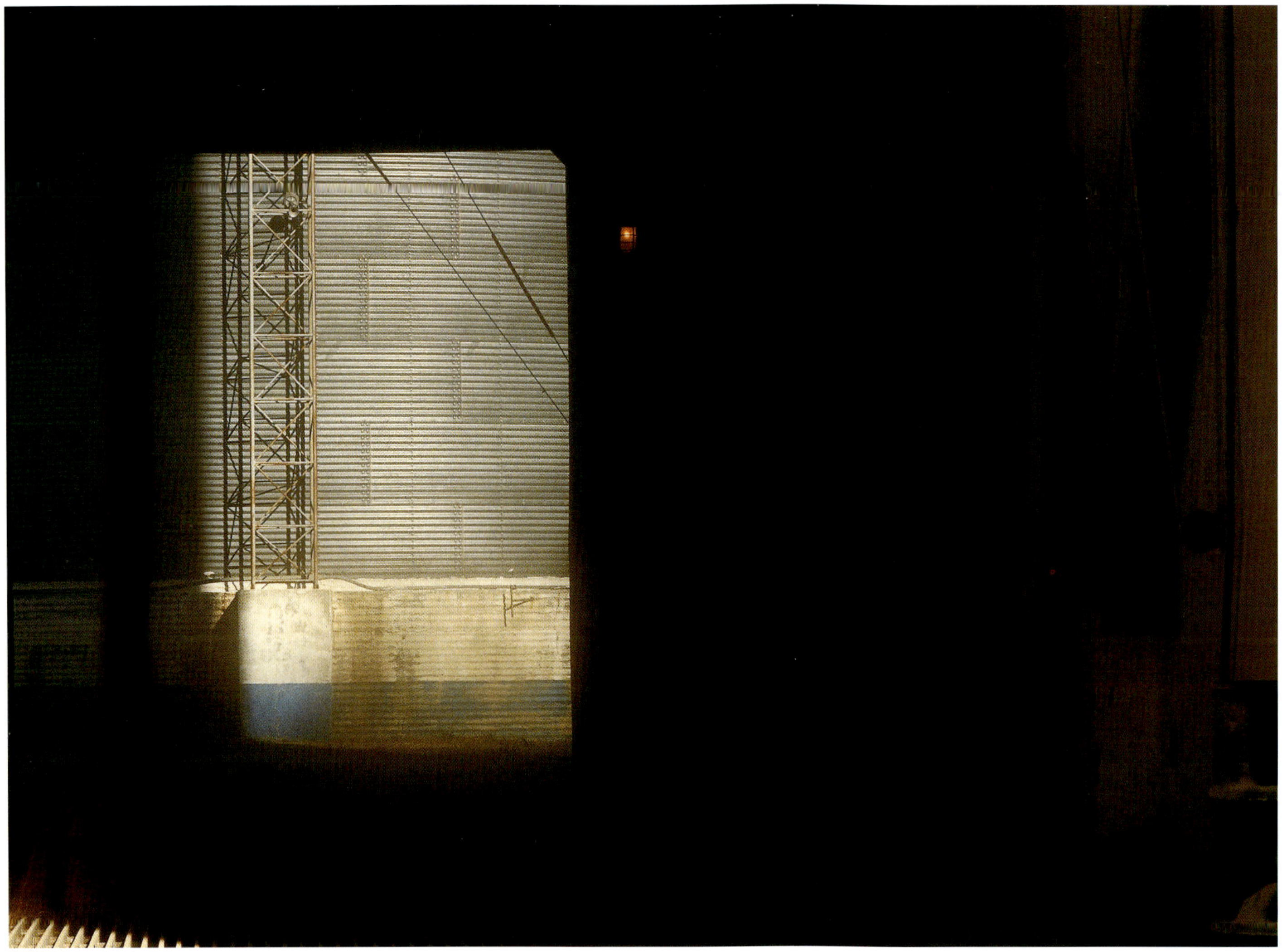

MILO DUST
Beeler, Ness County, Kansas

hard place, and that they survived because of family and neighbors. They know they owe their own people everything. But they don't owe much to outsiders. In Europe, for centuries, they had been downtrodden. This land, in America, was their chance at independence. They had to survive. A widely accepted story about the absolute value of family and community consequently lies at the heart of their regional mythology.

Kathleen Norris, in *Dakota: A Spiritual Geography*, tells us we'll never learn the real stories from histories that tell of a "progress from homesteading days to the present." The important stories, she says, are never spoken. Norris tells us to talk to the homesteading and early merchant generation if we want to "learn the truth about the web of close knit families that make up an isolated town on the Plains." Many of them, she says, "the women especially, had an intense love/hate relationship with the Plains." People who found they were living in an economic colony adopted "a world-view in which everything from the outside is suspect, while everything local, especially that which derives from the immediate family, is good. These are families that have an exceptionally difficult time dealing with conflict and change. Change means failure."

Generations of immigrant settlers took enormous risks to get to the Plains. Many children and grandchildren of those survivors seem dedicated to wrapping their good luck around them, cutting off any possibility of loss. Values are more and more commodified; personal worth is measured in terms of acreages and dollars. People come back to think of selfishness as a sur-

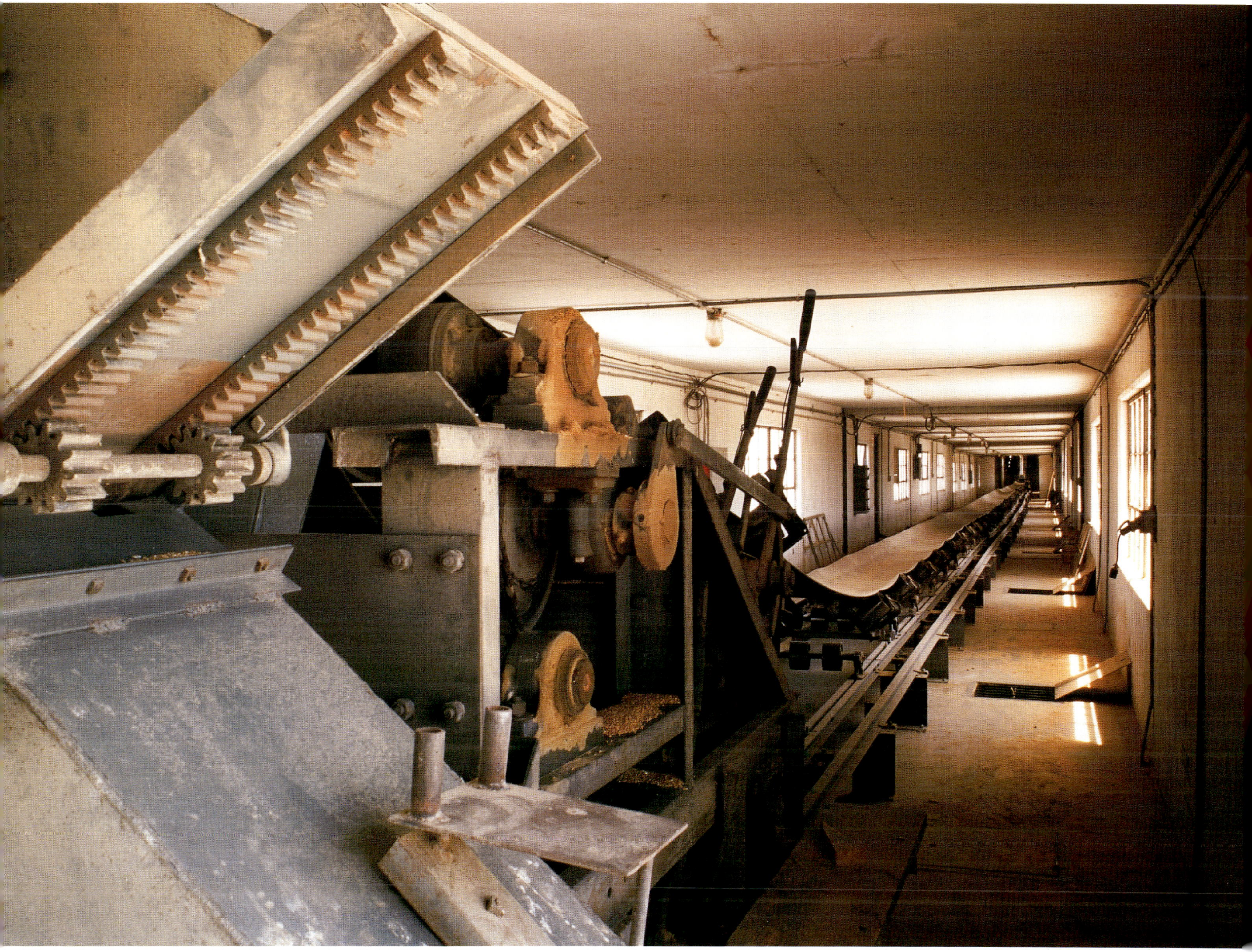

GRAIN CONVEYOR
Ness City, Ness County, Kansas

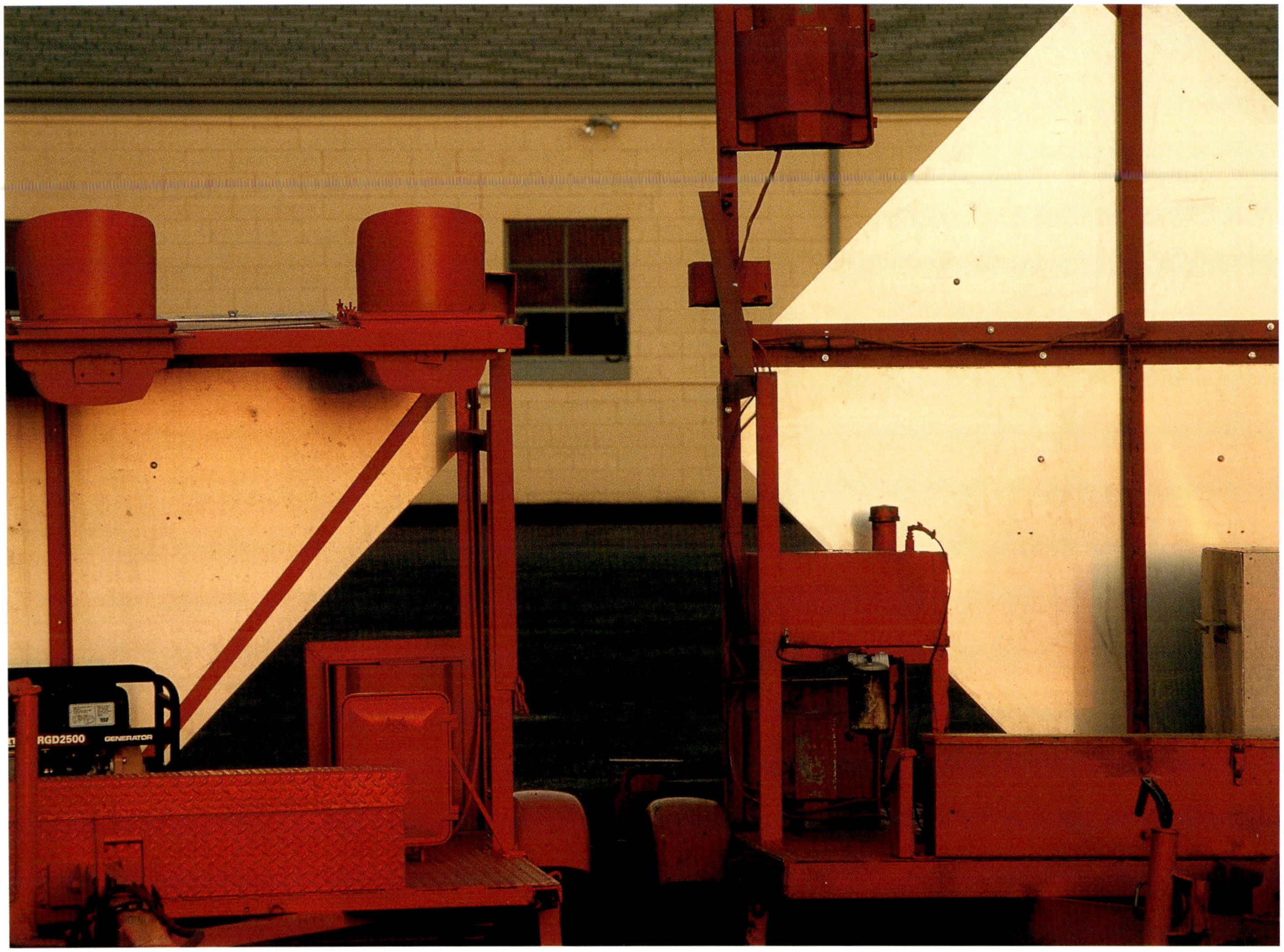

MAINTENANCE YARD
WaKeeney, Trego County, Kansas

vival art. Generosity toward anyone other than family or immediate community begins to be regarded as foolishness.

We yearn to hammer our advantages into place, but the world will not hold still, it will not be nailed down. The urge to edit family and cultural histories, to hide behind decorum and "make nice," is a defensive trait that runs strong in people who feel excluded from the adventures of the so-called "Great World," as do many people along the 100th meridian. For most people in the heartland, that great world — Paris and New York, Los Angeles and Hong Kong and Mexico City — is an unreal set of distant places they see on television, or in movies, or read about; events that transpire in those places don't seem to be particularly relevant to lives on the Great Plains.

The result, often, is isolationist purposes and politics, failure to attempt nourishing anyone outside family and immediate community, which in our increasingly interwoven world is a fundamental way to be irresponsible toward both the world and the self. We build psychic walls to protect all we revere, and oftentimes leave ourselves no choice but to lead defensive, humorless, and denuded lives.

Frivolity

Along the 100th meridian, in the Dakotas and the Sandhills and south of the Platte and on wheatgrounds in Kansas, in the red earth fields of the Oklahoma and Texas panhandle country, and in the Permian Basin and on the Edwards Plateau, I wanted to think the fearfulness I witnessed wasn't as pervasive as it seemed, that it was something I made up.

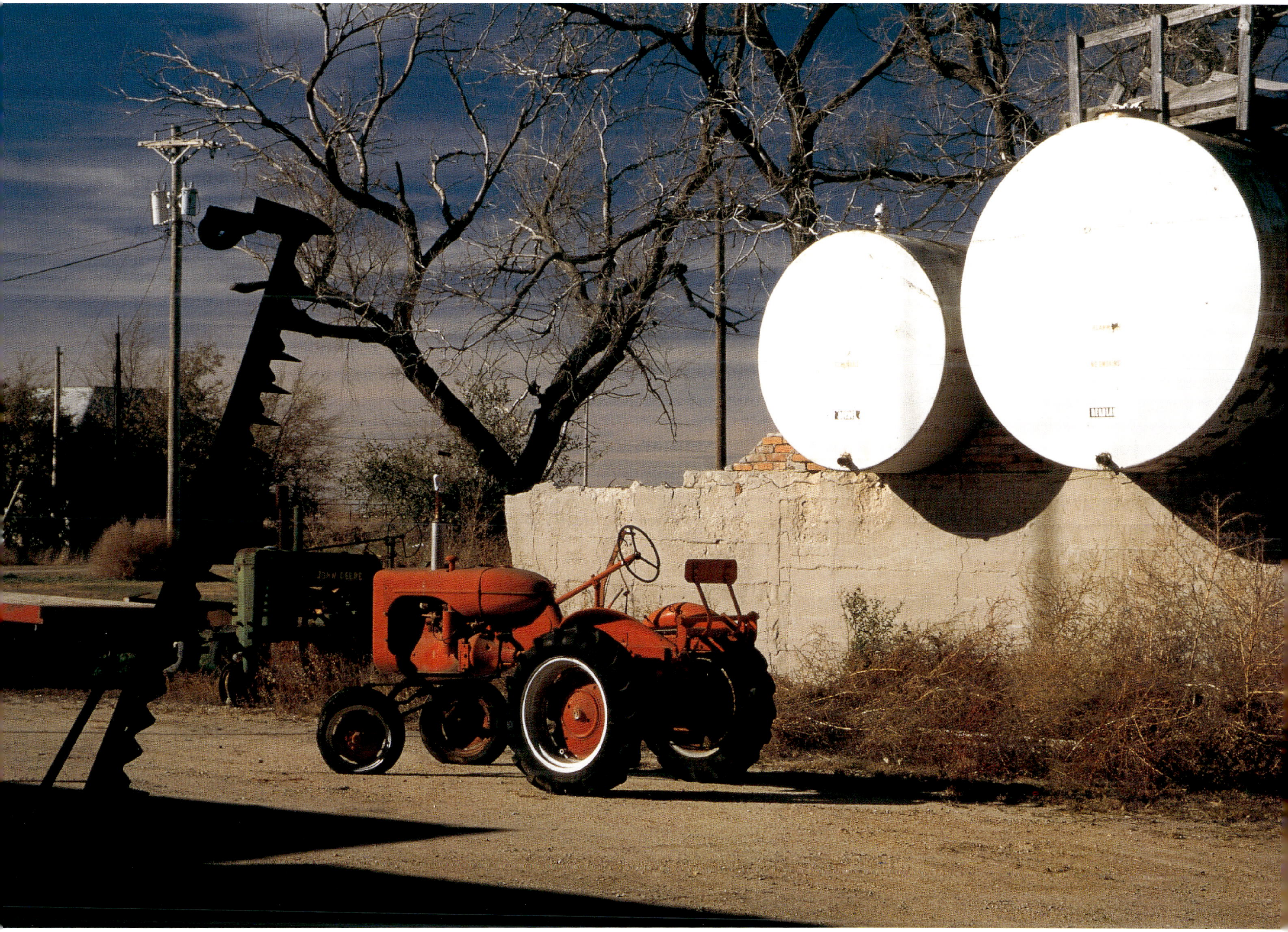

TRACTOR AND SILVER TANKS

WaKeeney, Trego County, Kansas

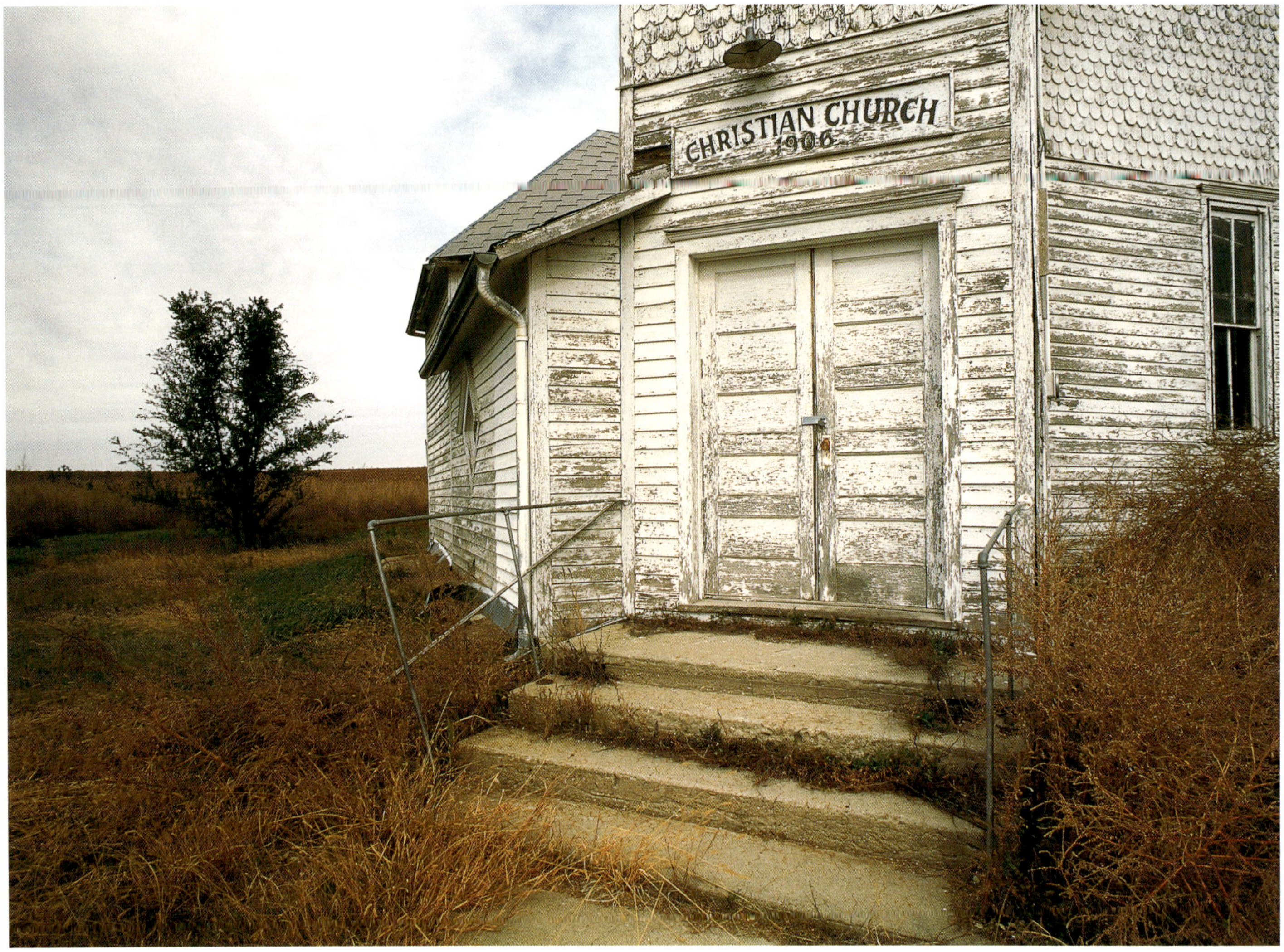

CHRISTIAN CHURCH: 1906
Edmond, Norton County, Kansas

What I missed was playfulness, which can be thought of as the saving grace.

This was borne in on me as I drove those Texas highways and listened to the collected works of Hank Williams, from "Hey, Good Lookin'" to "Your Cheating Heart," joy to sorrow. Blacks in the Deep South invented the blues, and yodeling Jimmy Rogers got white southerners tuned into the sound. But country music is a series of footnotes to Hank Williams, the genius of whiteman blues. "Why don't you love me like you used to do?"

Bob Wills, who grew up fifty miles west of the 100th meridian, in Turkey, Texas, invented western swing. He softened the irony running through Hank Williams with fiddles and yipping fun. Small-town bands in the West still play "Take Me Back to Tulsa" and "Steel Guitar Rag." Bob Wills honored a communal necessity — frivolity. "There'll Be No Crying Tonight."

The interior of the continent, bisected by the 100th meridian, is a stage where our society has performed, and still performs, giving shape to its purposes. From Canada to Mexico it is marked as though written upon — plow grounds, windbreaks, tight homes, vast commercial farms, an occasional sweet and prosperous town, children at play, and concrete grain elevators towering over empty villages. What I saw looked to be a society experiencing a failure of confidence, and thus increasingly afraid of wildness, living out a sad cautionary tale about the degree to which our lives will be diminished if ever entirely domesticated.

Heading home toward Montana, I passed an abandoned drive-in theater out on

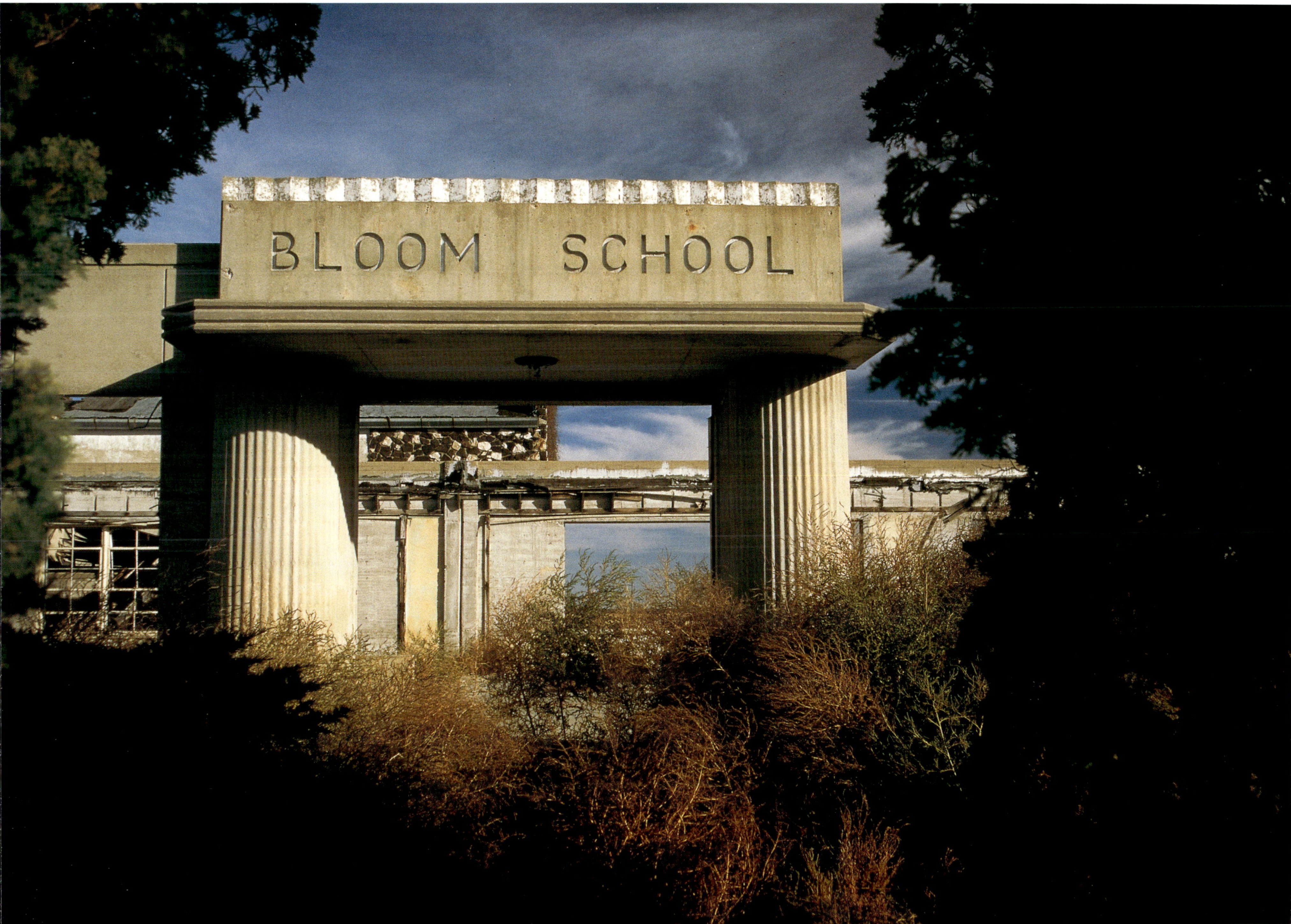

BLOOM SCHOOL
Bloom, Ford County, Kansas

BACKBOARD
Brownell, Ness
County, Kansas

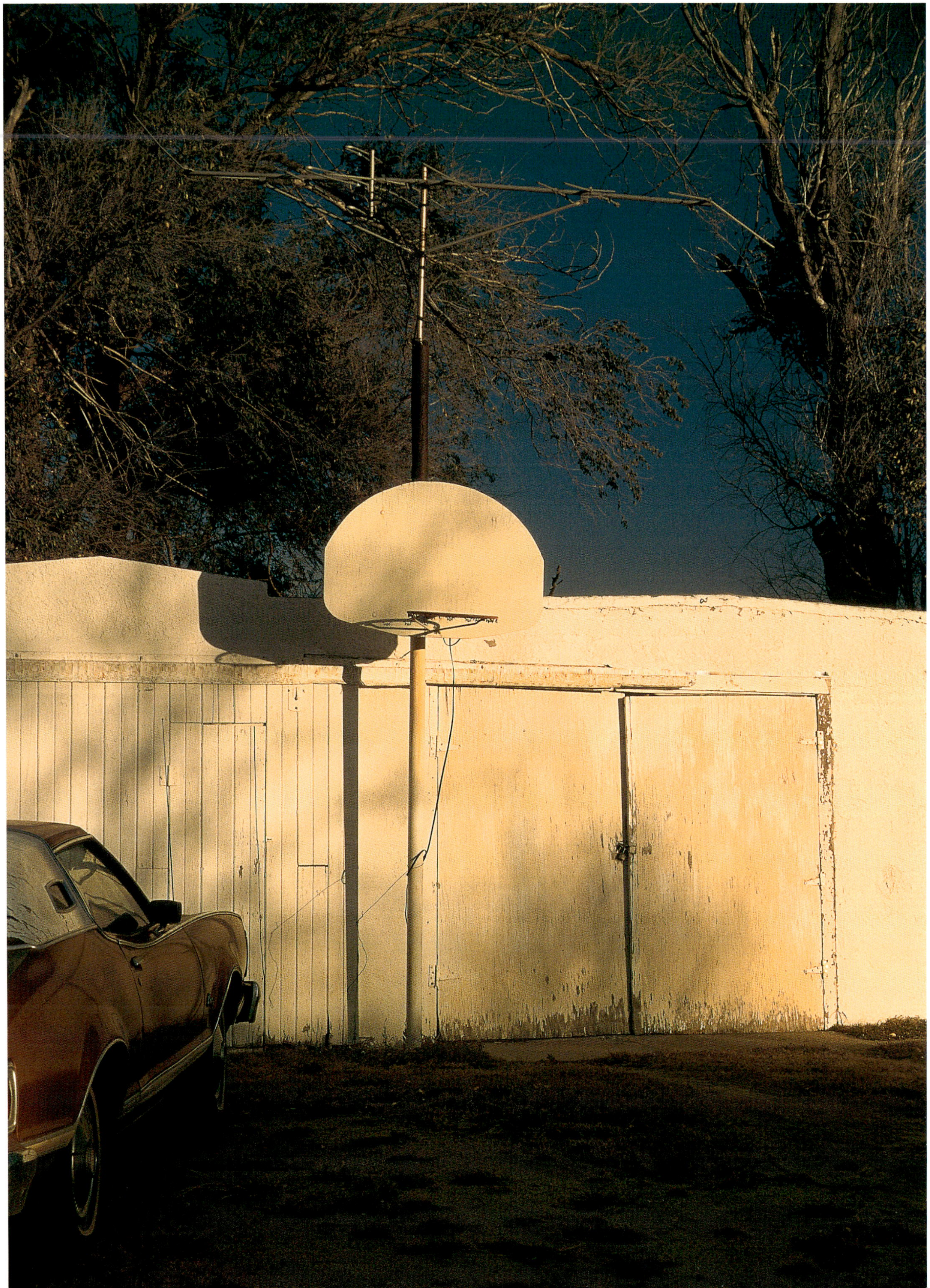

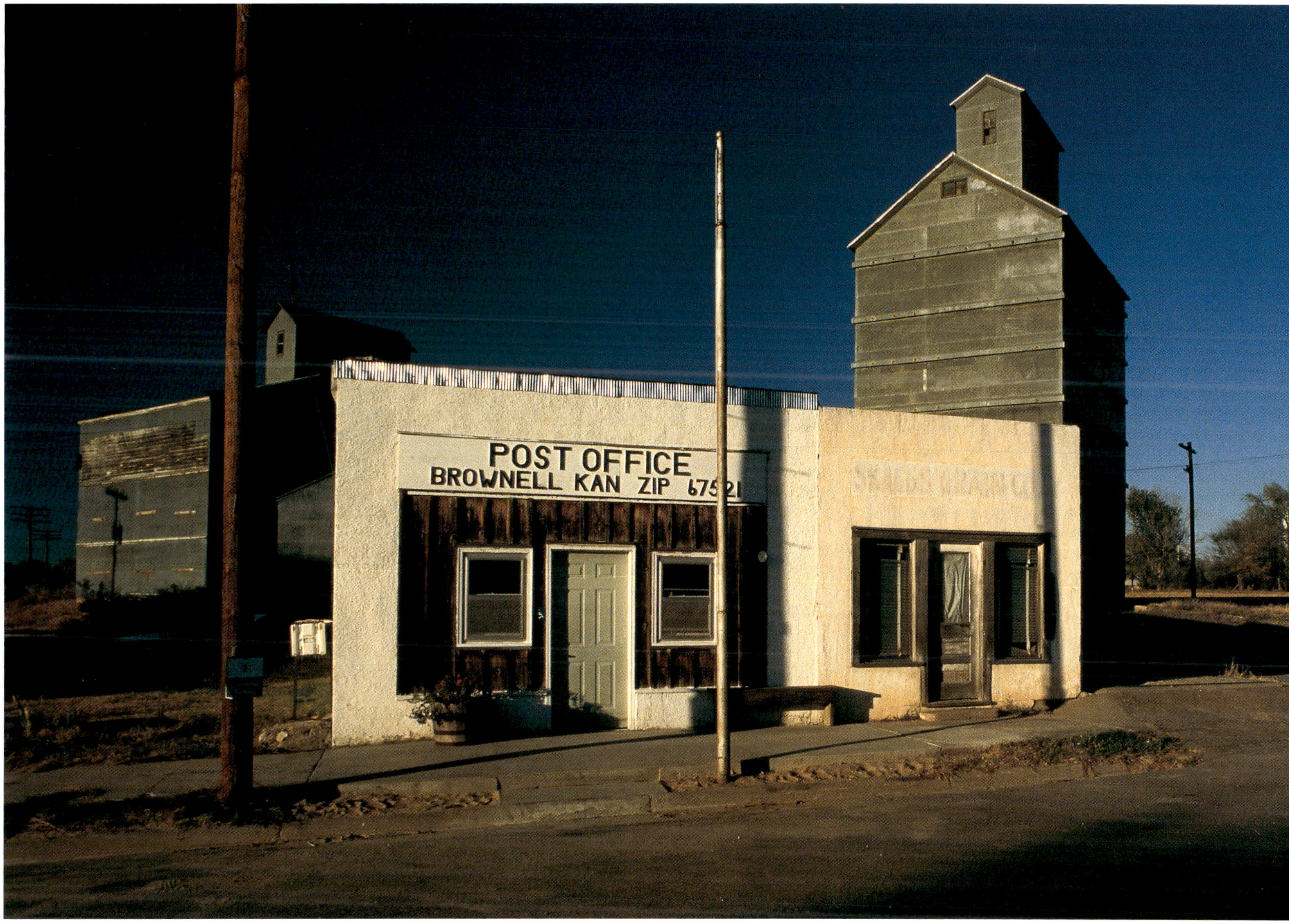

POST OFFICE
Brownell, Ness County, Kansas

the New Mexico plains east of Clayton. It looked like a falling-down apparition, a lost ship, sails whipped away on the sea of short grasses.

I was yearning for liberations, festivities in which cultural hierarchies are turned upside down, and fractured, in that way opening themselves to critique and reinvention, renewal and freshening. Rodeos and Friday night football games under the lights and horse races at the county fair and plain old nights on the town, anyplace where people are wisecracking and irreverent.

Later, in Clayton, I watched school children, some of them looking goofy with happiness, being driven around town in a big red fire truck. A bright soul had found a way to turn their day quick with possibility. As Mary Oliver has written, "like a million flowers on fire."

I yearned to go out on the edge of town and catch Bob Wills and his band in some old hall — sweaty dancing until daybreak, bad old boys and good old girlfriends, grinning and backslapping, a vast number of cold bottles of beer — and no remorse whatsoever.

Points of Departure

A Photographer's Perspective

Monte Hartman

It wasn't hiking weather in the Montana Rockies, even by the most macho outdoor standards. Like most visitors to Glacier National Park, my wife, Lee, and I were hunkered down under cover on the sagging, ancient bed, books on knees, for the storm's duration. Rain beat a white noise tattoo on our roof, on the Swiftcurrent complex, and against the entire eastern slope. Relaxed by repetitive noise, physical inaction, and warm sleeping bags, we oozed into the kind of edgy wakefulness that soon evaporates into afternoon sleep.

My attention had been focused on a Wallace Stegner essay that described the 100th meridian, that geographer's line down the middle of the continent, as the boundary between two different worlds, when the notion of a place where the East ends and the West begins tweaked my imagination, and the idea of photographing America's 100th meridian was born. Where was it? . . . What does it look like? . . . Has anyone photographed it? I was awake with curiosity.

This is the story of the project that emerged that day, along with a few impressions recorded in the logbooks of my twenty-thousand miles along the spine of America, where the conditions of life change, and, for Steinbeck, "where the map folds."

■

Openness begins at the meridian. The East is characterized by connectedness and proximity, farm, town, and city in a quilted urbanity. But at the meridian, the closeness and community of the East disappear,

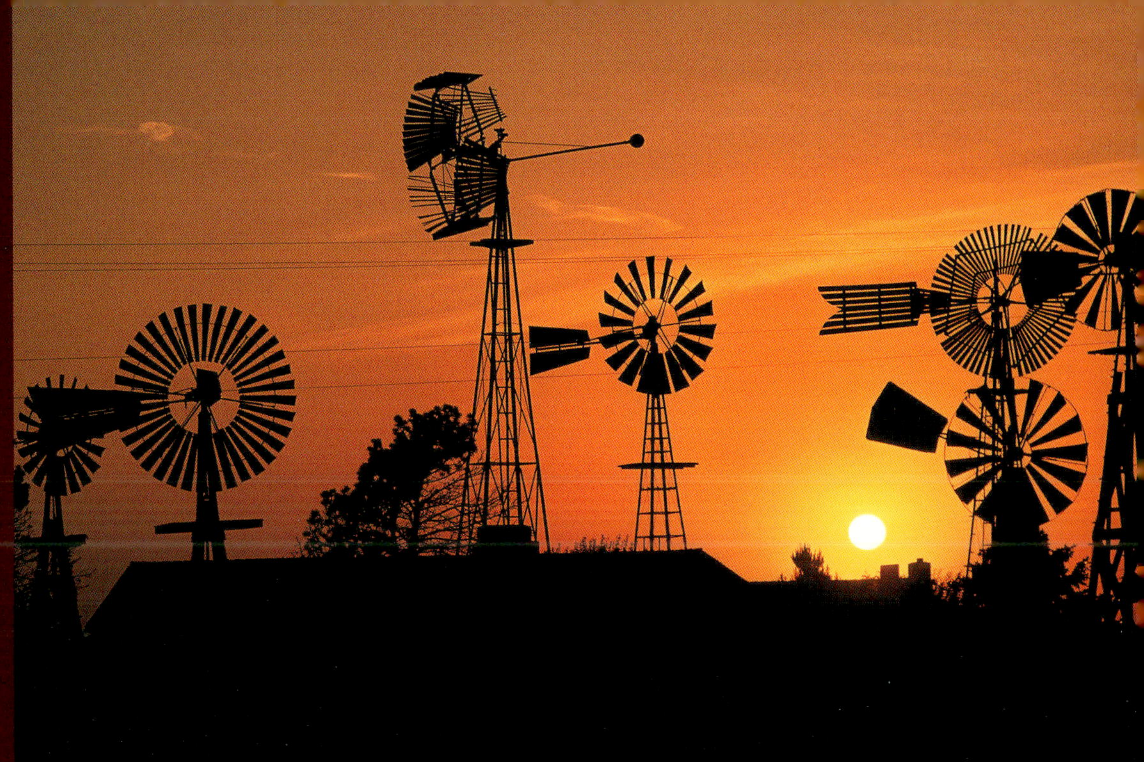

OKLAHOMA

WINDMILL MUSEUM

Shattuck, Roger Mills County, Oklahoma

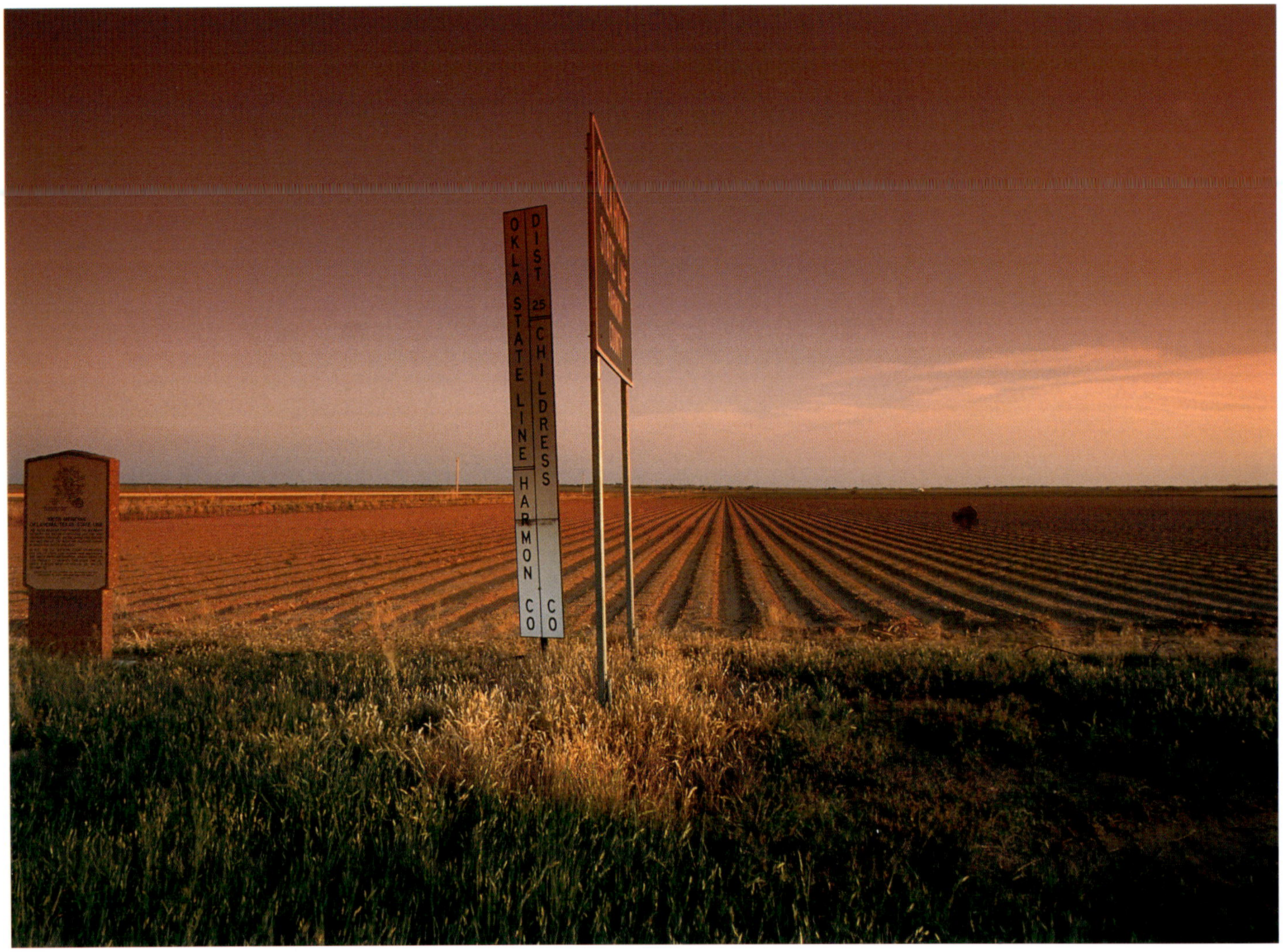

ON THE MERIDIAN
Harmon County, Oklahoma

replaced by the space of the West, where connections are tenuous and distance dominates the culture.

In this gigantic place, self-reliance was bred by selection, not mutation. It is not for the faint-hearted or the comfort-loving. Pioneers surged across the meridian, only to be beaten back repeatedly by drought and blizzard. In Stegner's words, "in the marginal zone between humid Midwest and arid West it was easy to be deluded, for the difference of just one inch of rainfall or a slight variation in the seasonal distribution would make the difference between success and failure." And there was much failure. The meridian was a no-man's land on which an heroic drama was enacted.

Even today, the landscape shows the physical scars of these struggles — the occasional detritus of a failed farm is accented each spring with hollyhocks emerging from the weeds and weathered boards, or by a rangy lilac bush, the sole survivor of a farm wife's optimism. Nearby, a shallow depression marks a collapsed vegetable cellar, and, a bit farther away, a drunken track of fence posts weakly insists on a now irrelevant separation of fields.

From Mexico, the 100th meridian enters the United States just west of Laredo. On its 1,900-miles from the Rio Grande to the Canadian border, the line crosses central Texas to form the Texas-Oklahoma border, crossing Oklahoma's panhandle and the central regions of Kansas, Nebraska, South and North Dakota. It passes directly through Dodge City, Kansas, Cozad, Nebraska, and the geographic center of North America near Rugby, North Dakota. In addition to the

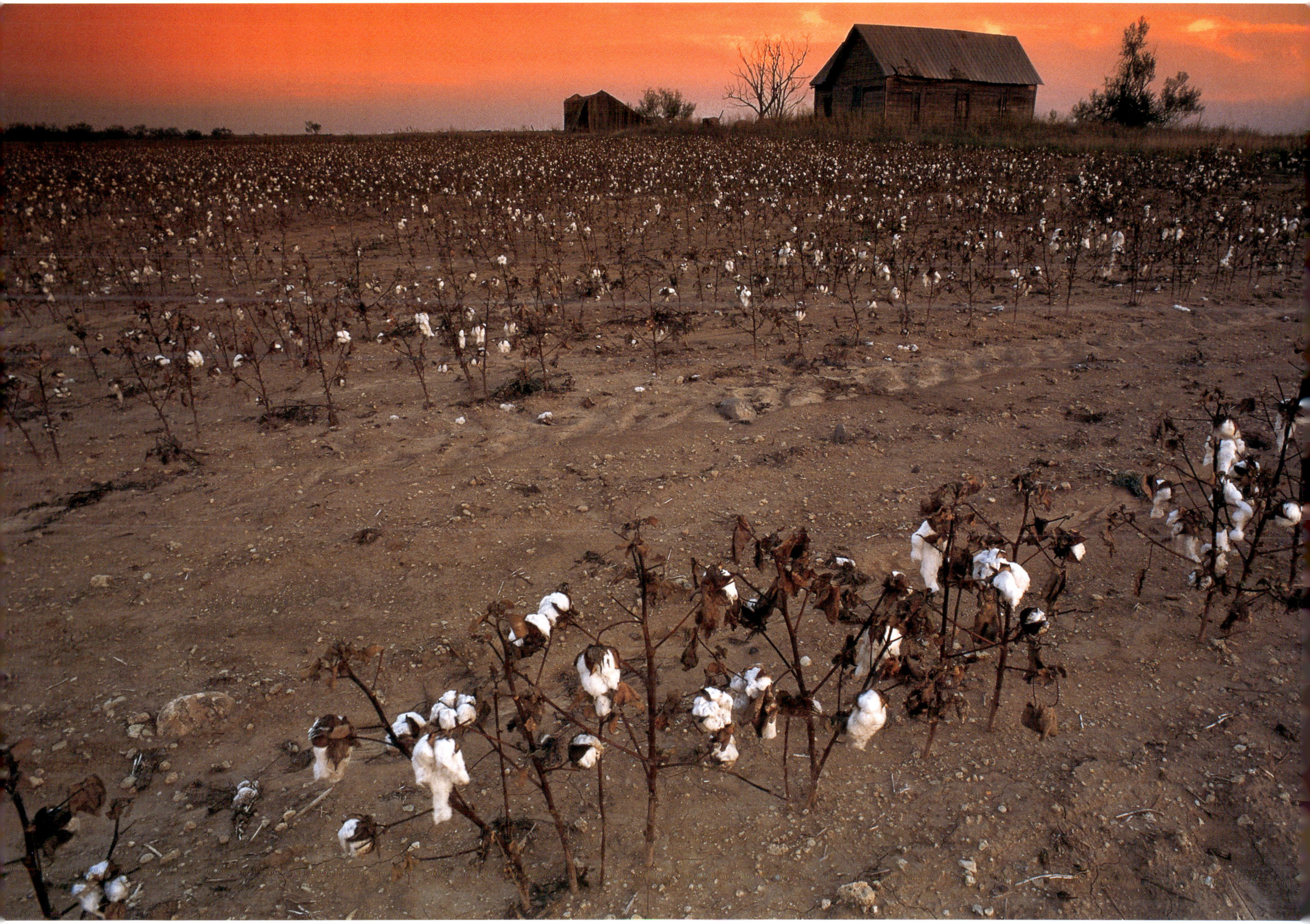

COTTONFIELD AT SUNRISE
North of Red River, Harmon County, Oklahoma

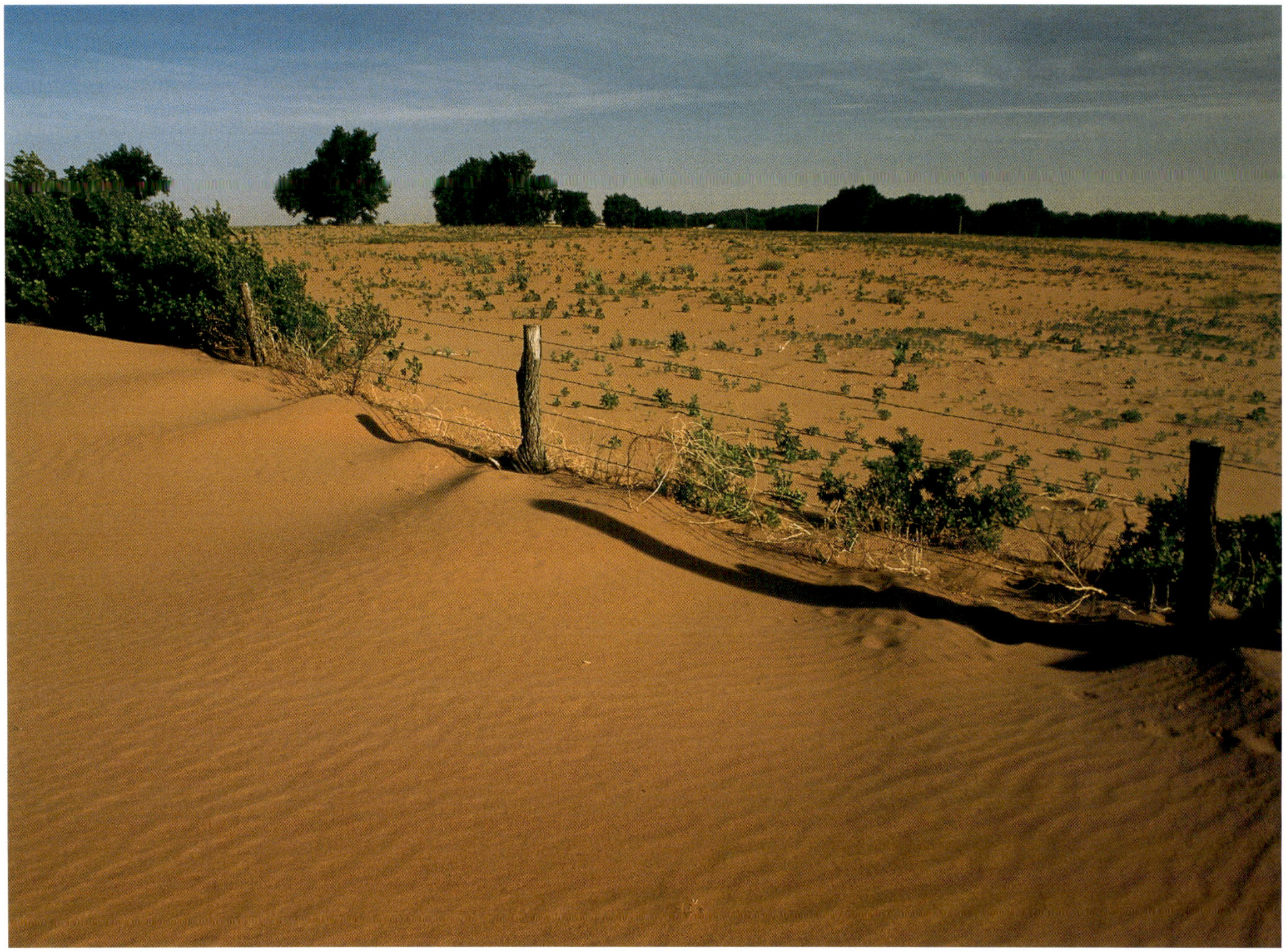

WINDBLOWN FIELD
East of Sweetwater,
Roger Mills County,
Oklahoma

Rio Grande, it crosses the Colorado, Red, Canadian, Arkansas, Platte, and Missouri rivers, touching fifty-six counties in six states, intersecting the track of the Lewis and Clark expedition, the Santa Fe Trail, and other historically important westward migration routes.

Intrigued with the idea of photographing this huge, remote area, so sparsely marked on the map and 1,600 miles from home, I wanted to see what I'd be up against in the originally-estimated three years it would take to do the project. In April 1993, Lee and I set off to explore the line: 100 degrees west longitude.

■

From Amarillo, Texas, we drove east to Childress, crossing the meridian in the predawn Red River mist, filaments of fog clinging to the bottoms and orange sunlight rising over the early shadowed land. We stopped at a 1920s–style bungalow, mud-spattered red at the base, shading upward into orange and then yellow. The house seemed to emerge from red muck and silvery mist, and I made the first photographs for the project there.

As we drove north from Childress, our fascination with the red earth was accompanied by a sense of emptiness. In this old cotton country, we found one remote town after another, each smaller and more deserted than the previous one. The number of boarded-up businesses was startling. Plywood-faced storefront after storefront formed continuous horizontal stripes, interrupted here and there by a VFW hall, a senior center, or a tiny cafe. When we left

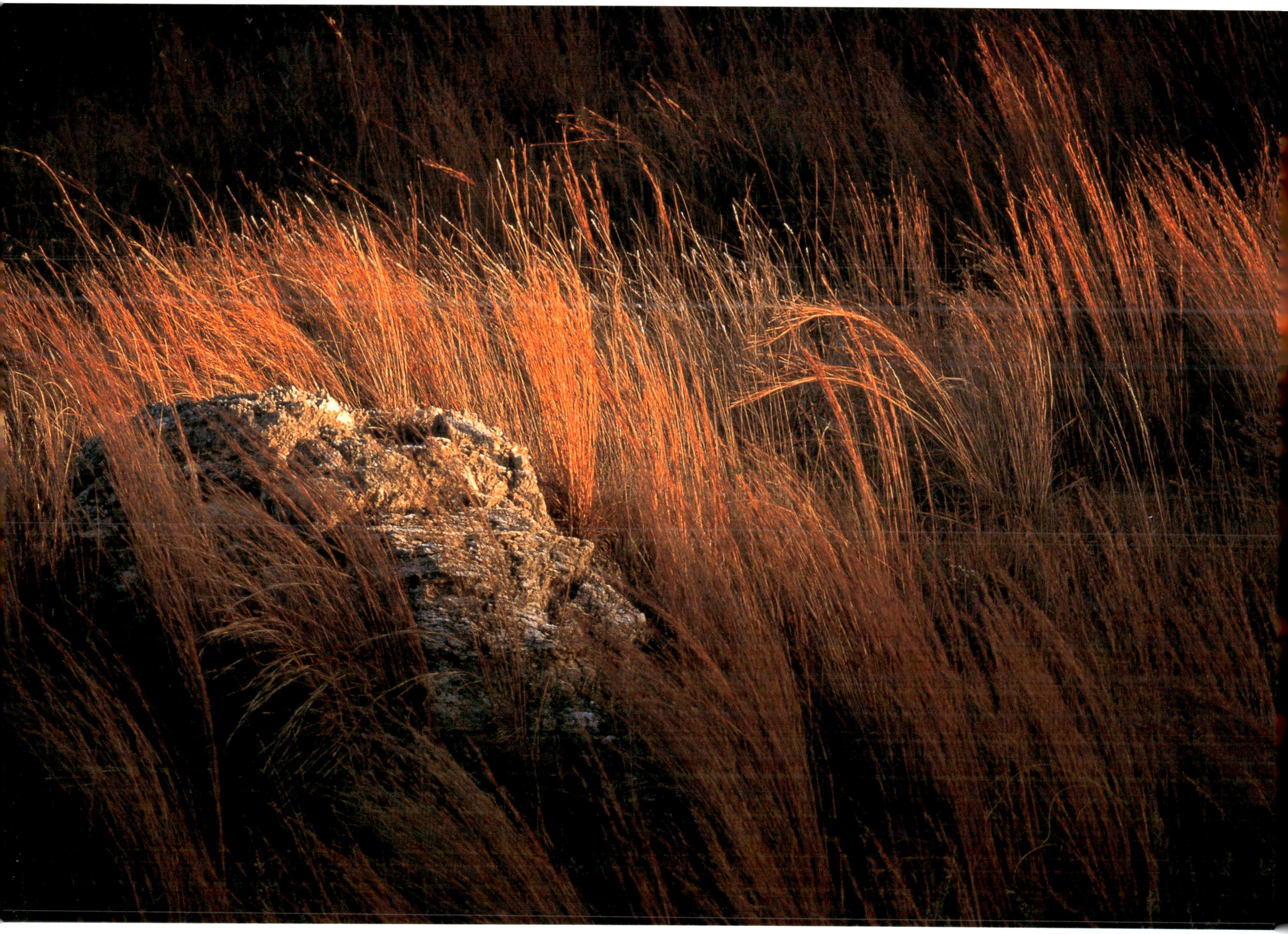

ROCK WITH FALL GRASSES
Elm Fork, Red River, Oklahoma

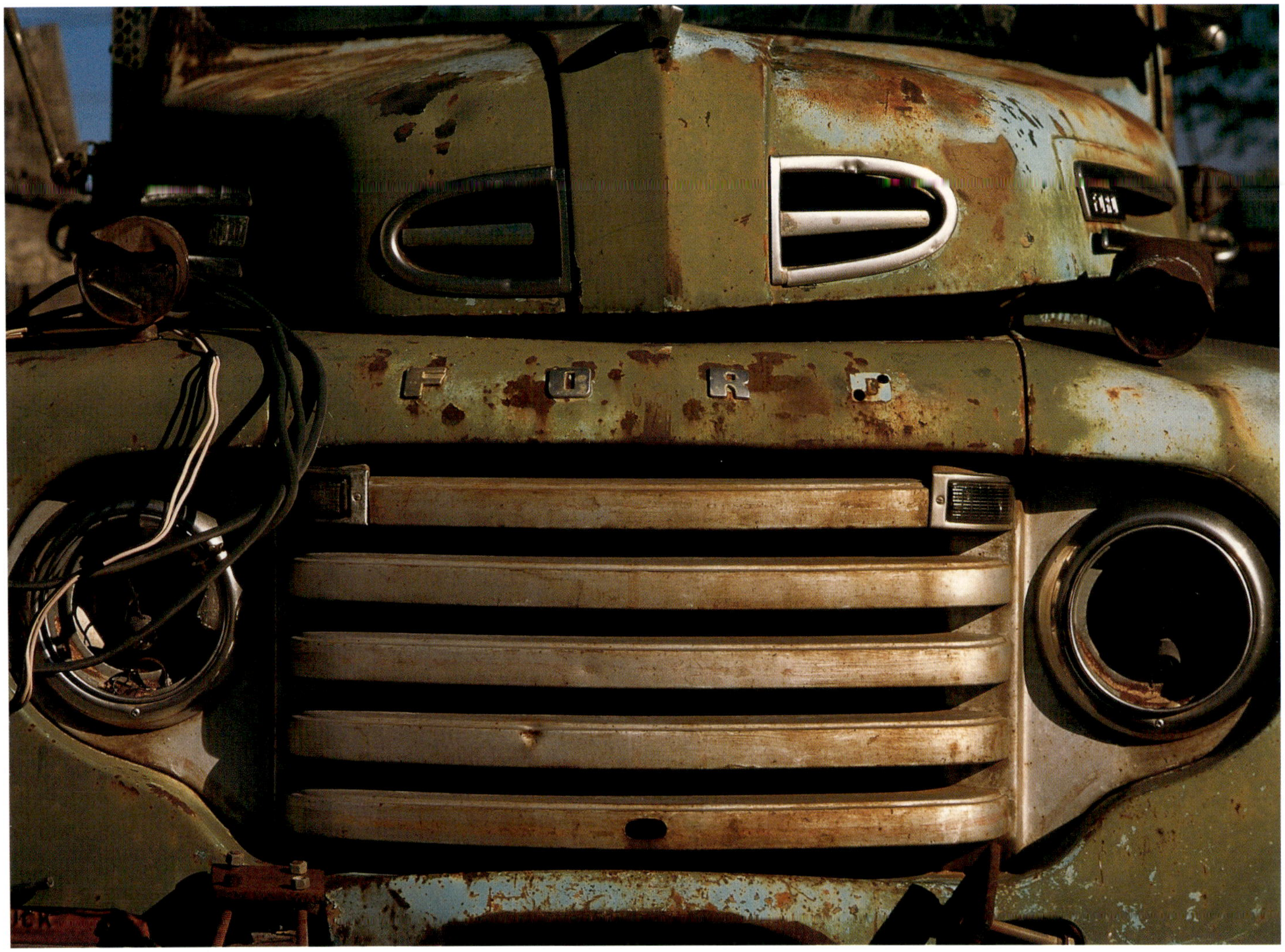

FORD TRUCK
Durham, Roger Mills County, Oklahoma

these places for the openness of the countryside, our sense of foreboding diminished, only to reemerge at the next town.

Meandering north across the nested borders of Oklahoma and Texas, we entered Kansas feeling gloomy. At the main intersection in Englewood, Kansas, I stopped to scan a depressing scene of deserted gas stations and decaying buildings. It recalled the surreal mood of postwar annihilation films about the earthly survivors of a nuclear Armageddon.

Exploring Englewood with my cameras, I almost stumbled into a gaping excavation, indecently exposing a brown sewer pipe unearthed for some long-forgotten reason, but never reburied. Remnants of furnishings and appliances spilled out of broken buildings. Overhead, wires dangled from empty streetlights. Gray squares of sidewalk, shrunken into islands, became stepping stones in a sea of waist-high weeds. It seemed as though I could hear the voices of past residents whispering their aborted dreams, their hopes for full lives in a land that had promised so much to so many.

Across the road, I photographed the municipal swimming pool, a crumbling blue-green metaphor. The children were gone. A 1940s vintage Chevvy was stored under a ragged tarp at poolside, optimistic, as if ready to start up whenever children and their laughing voices returned to claim their pool.

As I drove northward, I again heard voices drifting across the plains from abandoned homesteads. The meridian lands and hollow hamlets were echoing the humanity stripped from once-thriving communities.

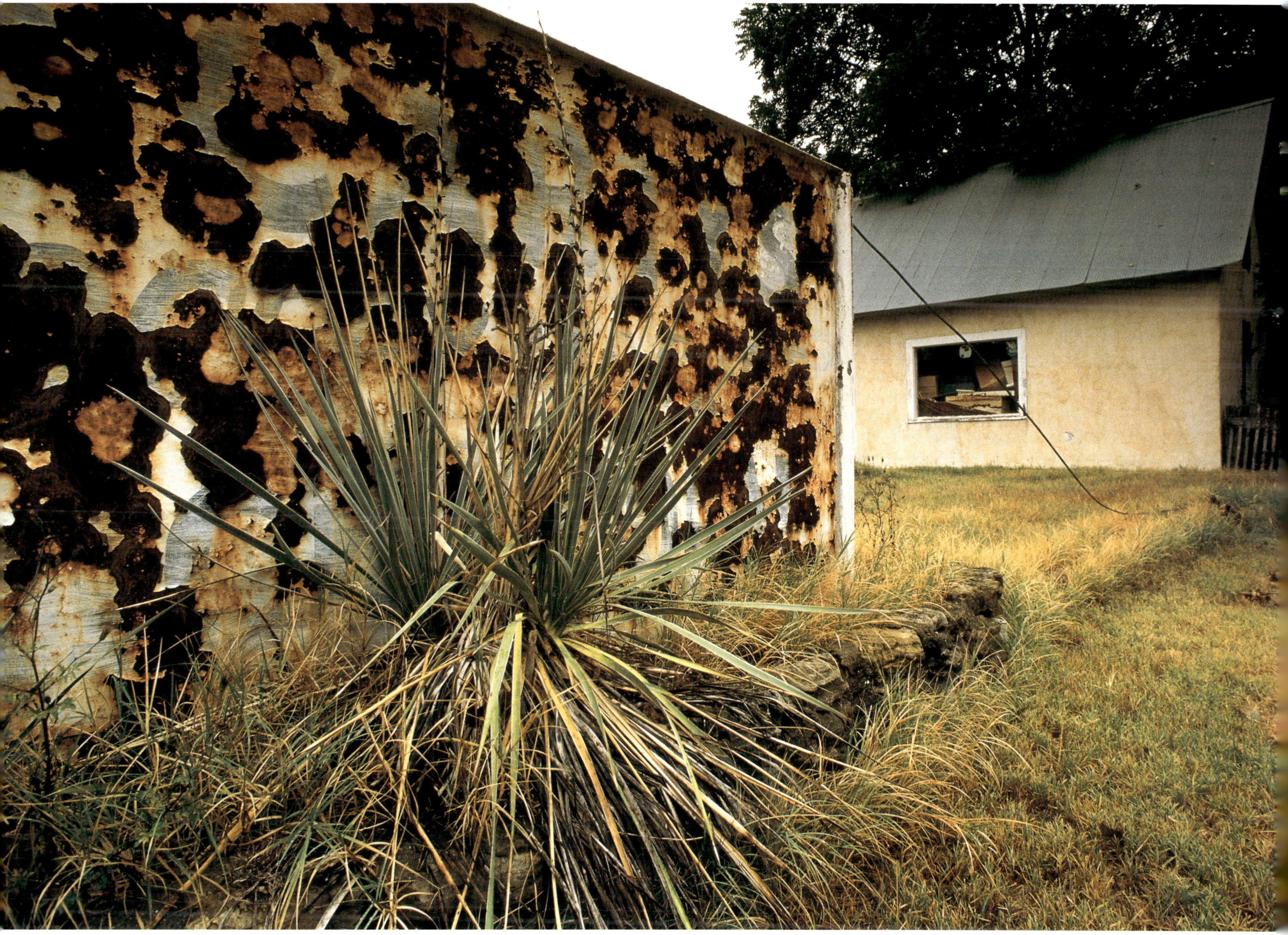

YUCCA
Erick, Beckham County, Oklahoma

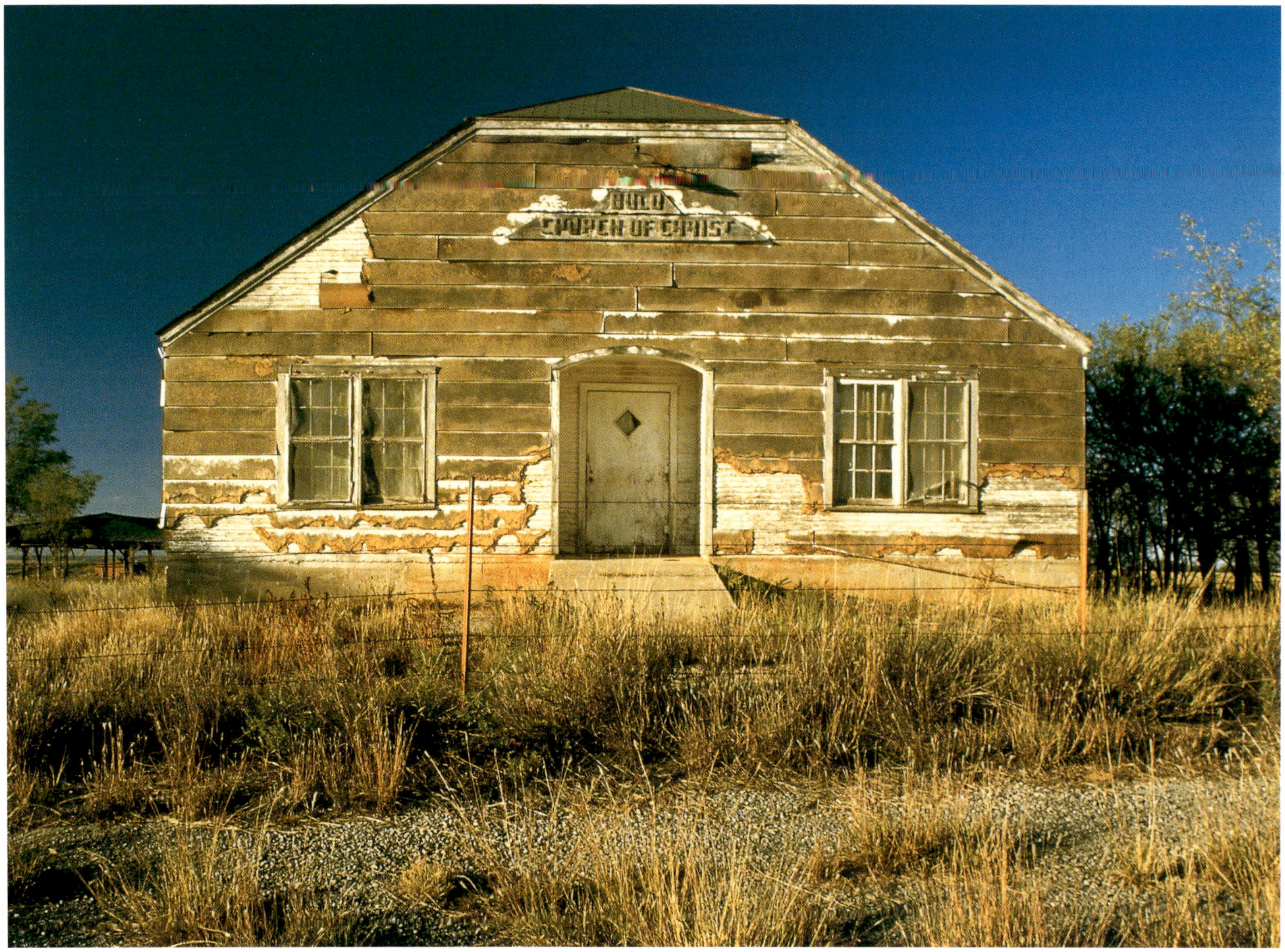

BULOW CHURCH
Prentiss, Beckham County, Oklahoma

In acknowledgment of those ghostly voices, I was totally committed to doing the project when we left Englewood.

■

Our exploratory trip took us up the meridian to Canada, down through the Nebraska Sandhills, and west on Interstate 80 back to California. Brooklyn-born, I was a person of both coasts, having lived in the Golden State for almost fifty years; an old-time easterner and nouveau westerner. The project fit this dual identity, dovetailing with a new-found focus in my photography. Recently retired from another career, a shifting of my photographic attentions from shooting "pretty places" had begun during my first sunrise at Bryce Canyon, where I encountered more photographers in one place than I'd met in my entire life. Seeing the tangled, overlapping tripod legs, I realized that my photography needed to transcend "ultimate" images of beautiful places. Although photographic beauty sells calendars and picture books, perfect images have a kind of unreality, their focus on exotic places often unreachable to most of us, instead of the familiar places where we live and work.

New possibilities emerged as my photographic attention shifted to the anonymous landscape. I found that ordinary places — abandoned houses, well heads, grain elevators, and storefronts — could each manifest a unique history, a private dream, an echo of the energy and spirit of its builders. These are the weather-beaten, time-textured visual substance of the meridian drama. The 100th Meridian Project seemed ideal.

The project was first envisioned as a simple picture book, but after our initial

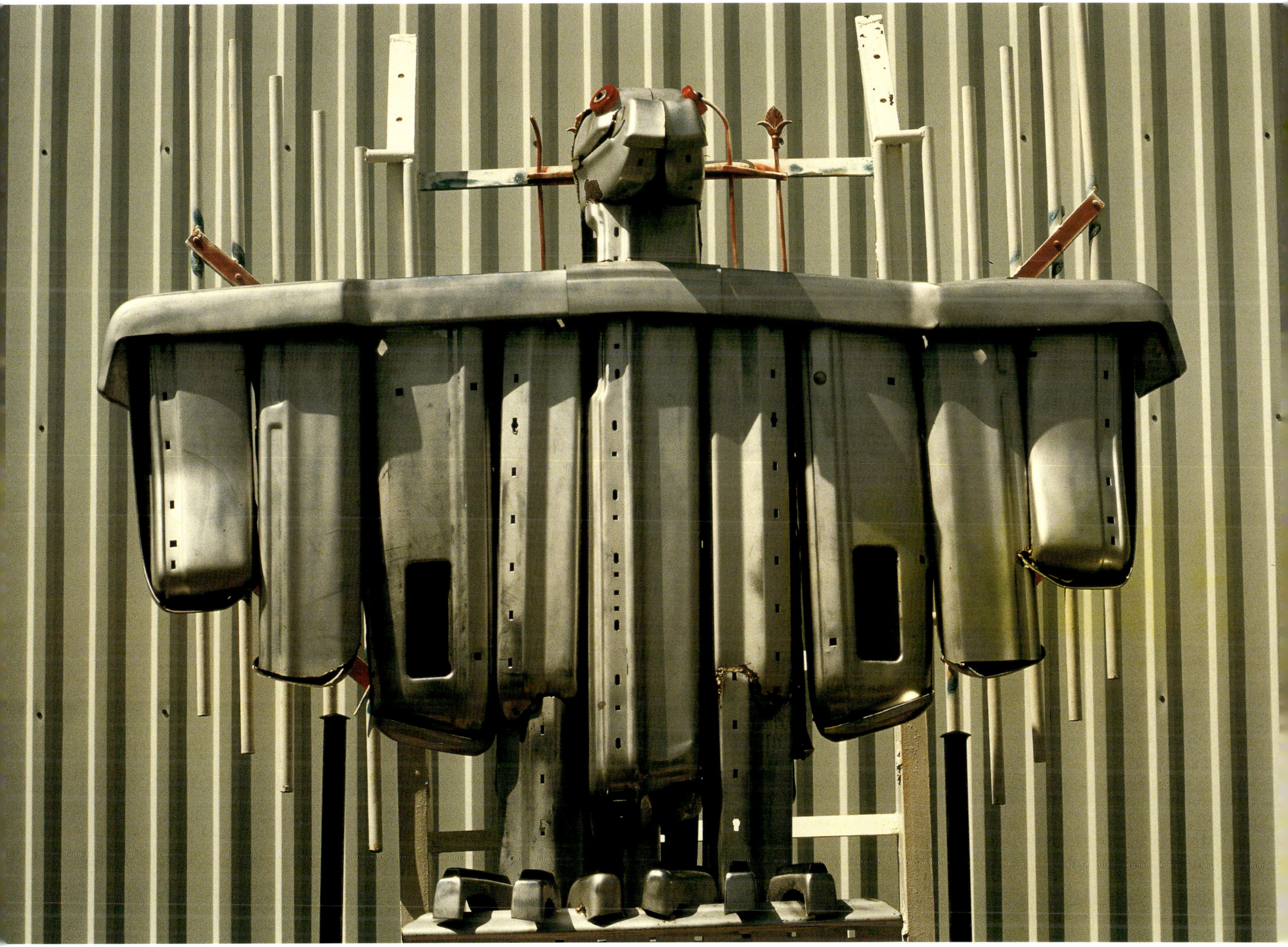

AMERICAN EAGLE SCULPTURE
Gage, Ellis County, Oklahoma

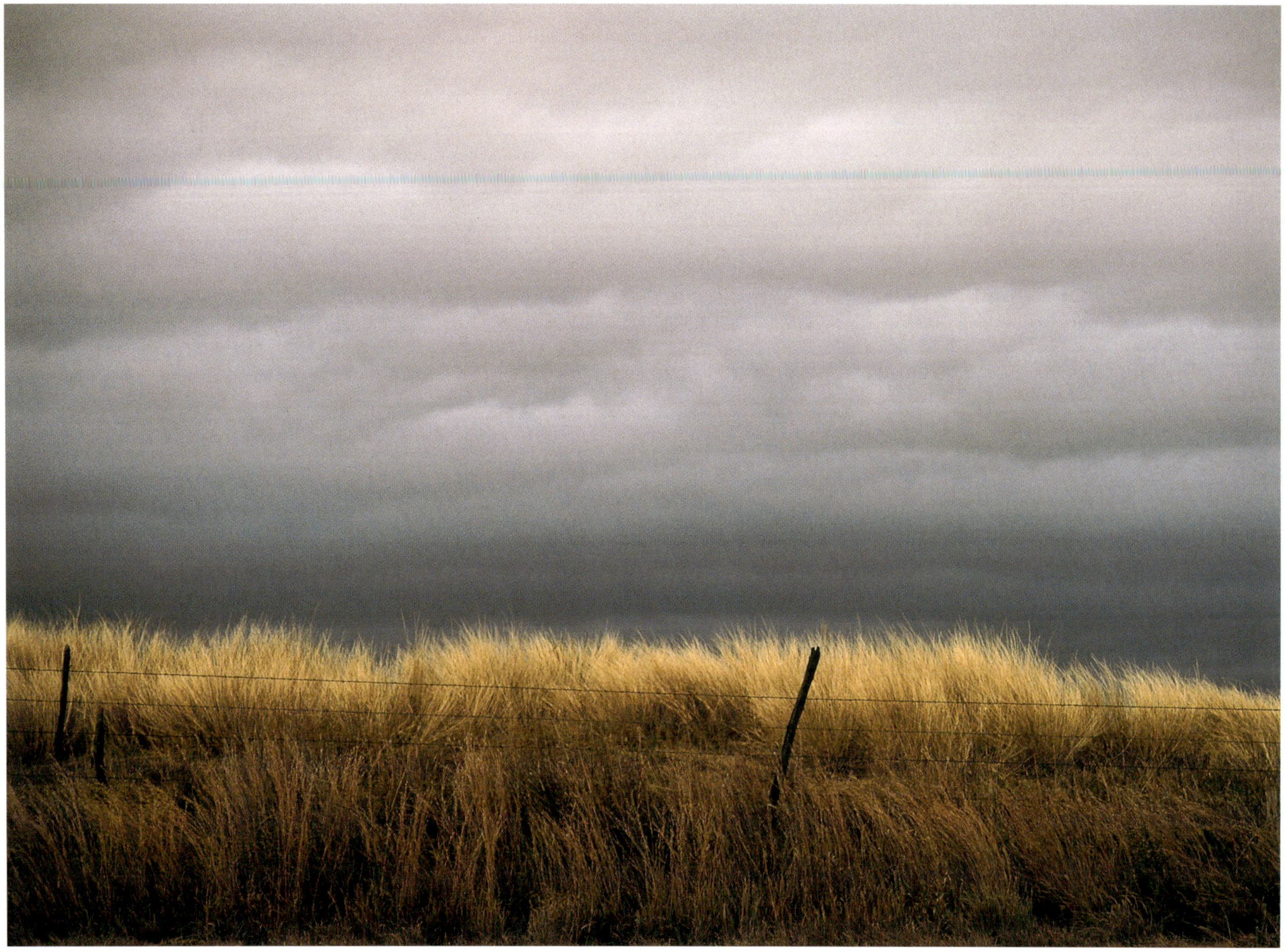

OVERCAST SKY
Near Crawford,
Roger Mills County,
Oklahoma

exploration, it was clear that a story was waiting to be told, and the enterprise evolved from a simple collection of images into a telling of the meridian human story in a meaningful way.

Matters of Geography

During the Ice Age, a great glacier dipped below the Canadian border at the 100th meridian. Its subsequent retreat left northern North Dakota with a moraine-strewn landscape of low, undulating hills speckled with lakes and ponds, the remnants of huge melting ice blocks. At this northernmost stretch, most of the land is treed, the occasional cleared fields ringed by boulders and marked with cairns of partially rounded rocks. At about midstate, the land flattens into open plains that extend southward through Nebraska and Kansas to the Oklahoma border, trees and shrubs lining streams and the low, rolling hills or "breaks" associated with major river courses: the Missouri, and to a lesser extent, the White, Platte, Arkansas, Canadian, and Red.

Below the South Dakota-Nebraska line, the meridian crosses the eastern margin of a bed of ancient sand dunes occupying substantial portions of west central Nebraska. Covering 19,300 square miles in a formation extending 265 miles east-west and 130 miles north-south, the Sandhills is the largest collection of dunes in the western hemisphere. Now covered with high grass, the dunes support a productive cattle industry and comprise some of the loveliest country on the meridian.

The topography south of the Sandhills is consistently flat and semiarid until the

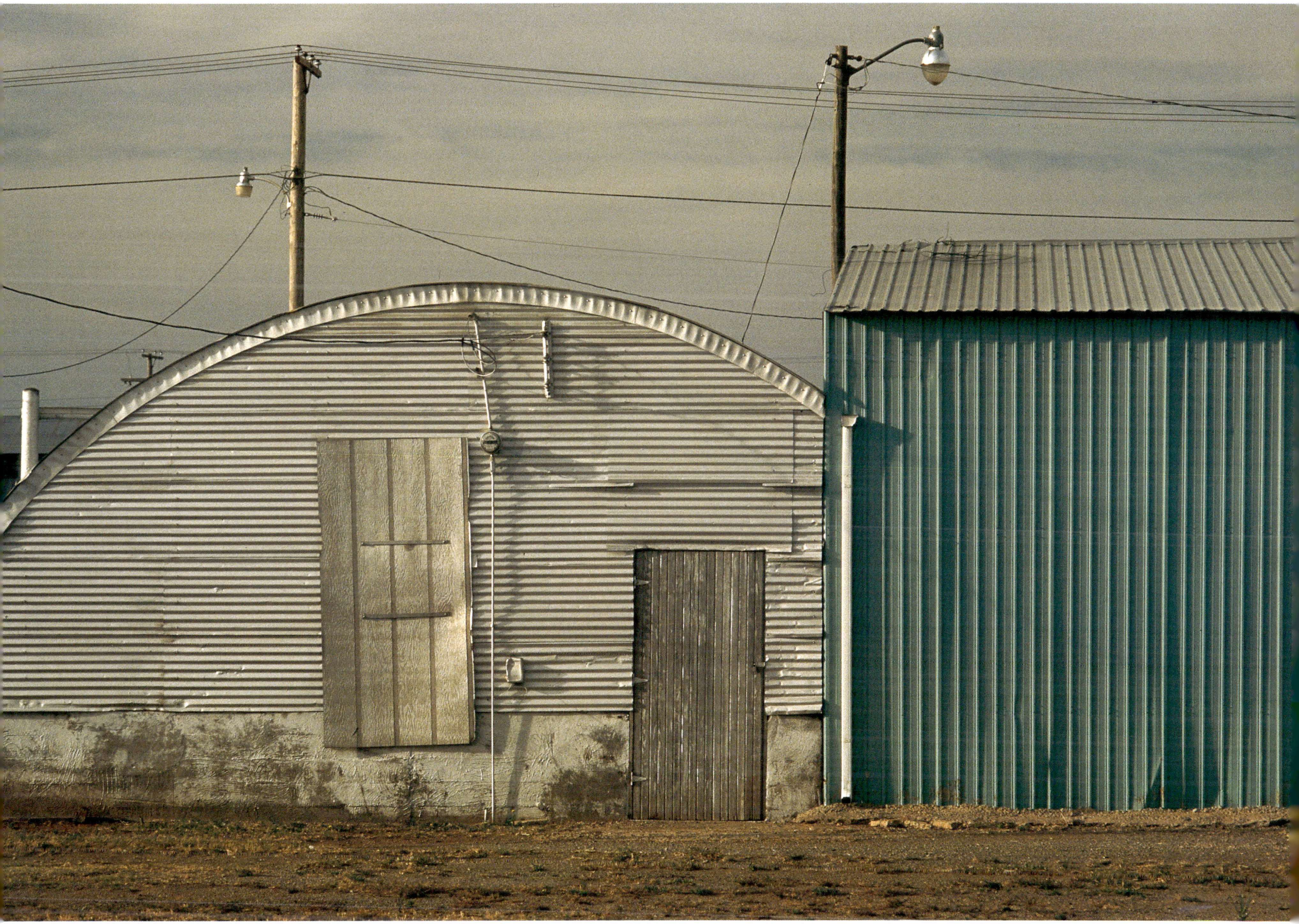

QUONSET COMPOSITION
Sayre, Beckham County, Oklahoma

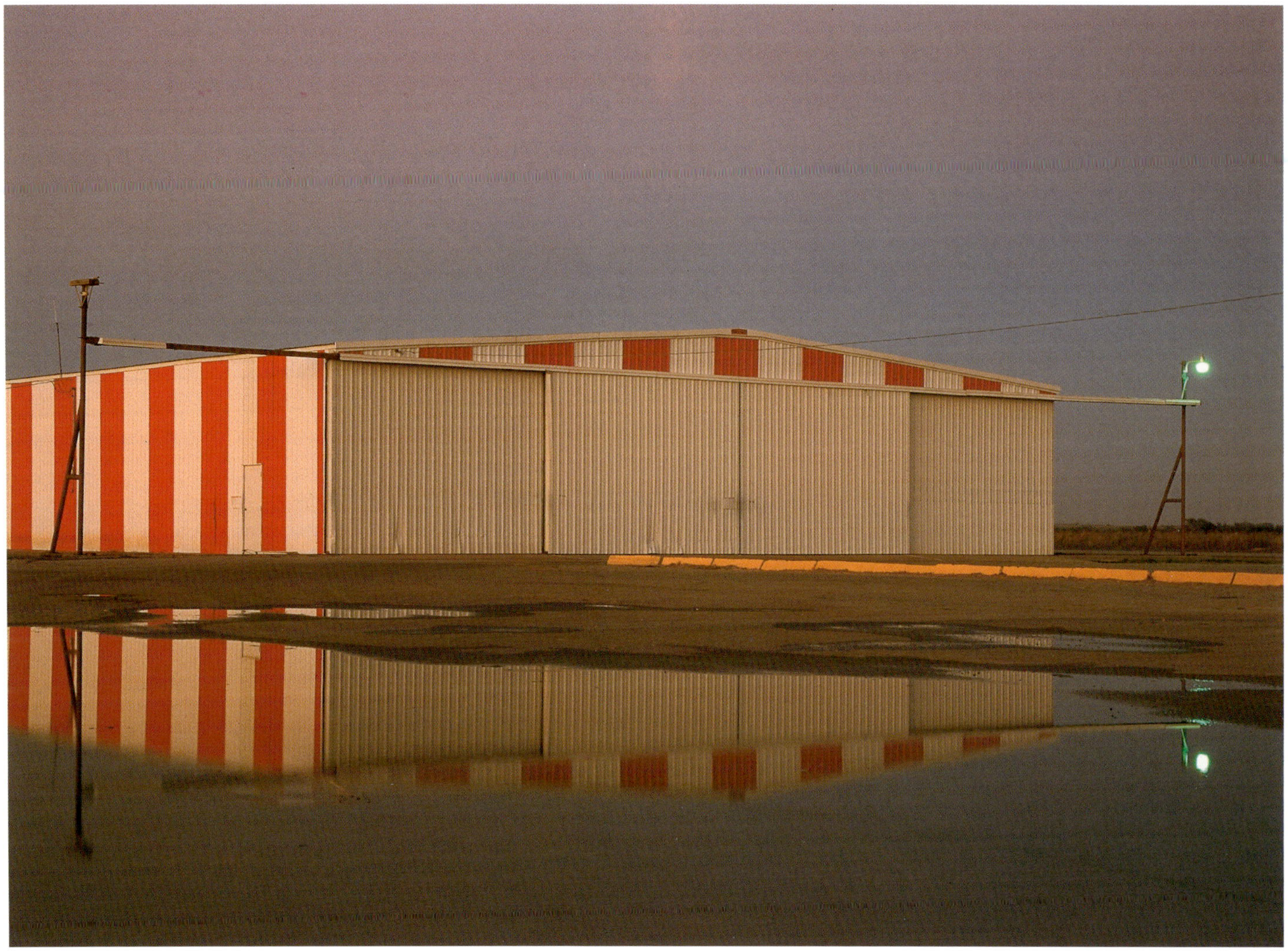

GRAY AG AIR
Hollis, Harmon County, Oklahoma

Kansas-Oklahoma line where evaporation increases and the land becomes hotter and drier. The appearance of sagebrush in southern Kansas, about twenty miles north of the Oklahoma border, begins a transition to a more consistently arid terrain, although extensive patches of grassland can still be found in southern Kansas and parts of Oklahoma.

The transition to a mesquite, juniper, and bunch grass ecology terminates at Erick, Oklahoma, and the dry landscape continues past Abilene, Texas, to the Texas Hill Country, interrupted only by the brick-red earth of the Red River at the Oklahoma-Texas line.

The Hill Country is an extensive formation of juniper- and pine-covered limestone ridges characterized by sculptured gray bluffs and canyons wildly different from anything else on the meridian, and certainly some of its most dramatic country. In the southern hills, the tree line plays tag with the meridian as the elevation changes, so juniper, pine, and mesquite alternate frequently with lush vegetation, oaks, and pecan trees. In Rio Frio, Texas, I saw the world's largest specimen of the eastern oak *Quercus virginiana,* var. fusiformis, a spectacular specimen more than 275 feet in circumference at the base, its branches shading a circle more than 100 feet in diameter. Not far away, on the dry side of the tree line, three-strand cattle fences give way to the five-strand wire of sheep and goat ranches.

At Crystal City, Texas, the Hill Country landscape flattens into the high evaporation, low humidity, and heat of the Rio Grande

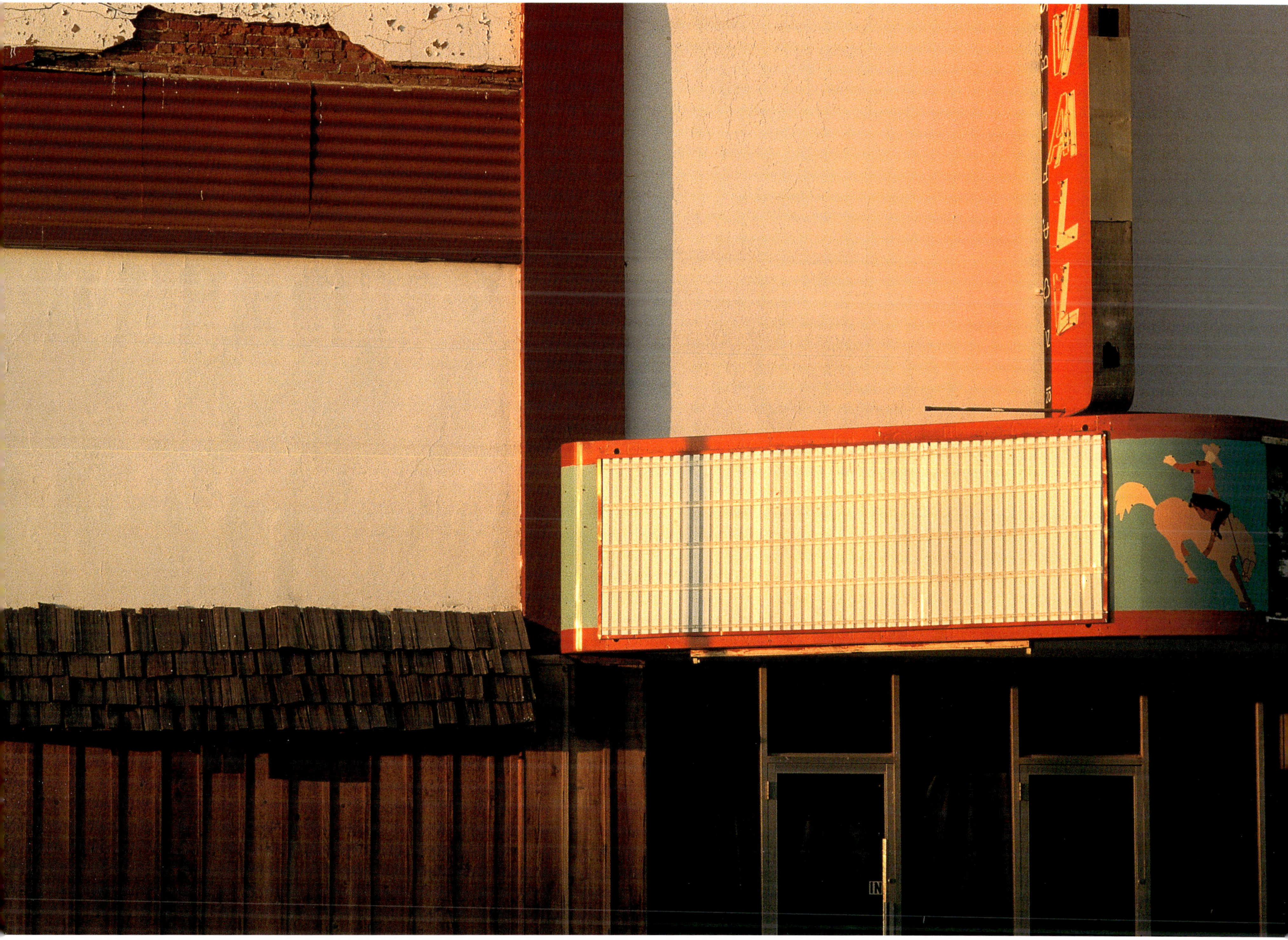

STOVALL THEATER
Sayre, Beckham County, Oklahoma

O K L A H O M A

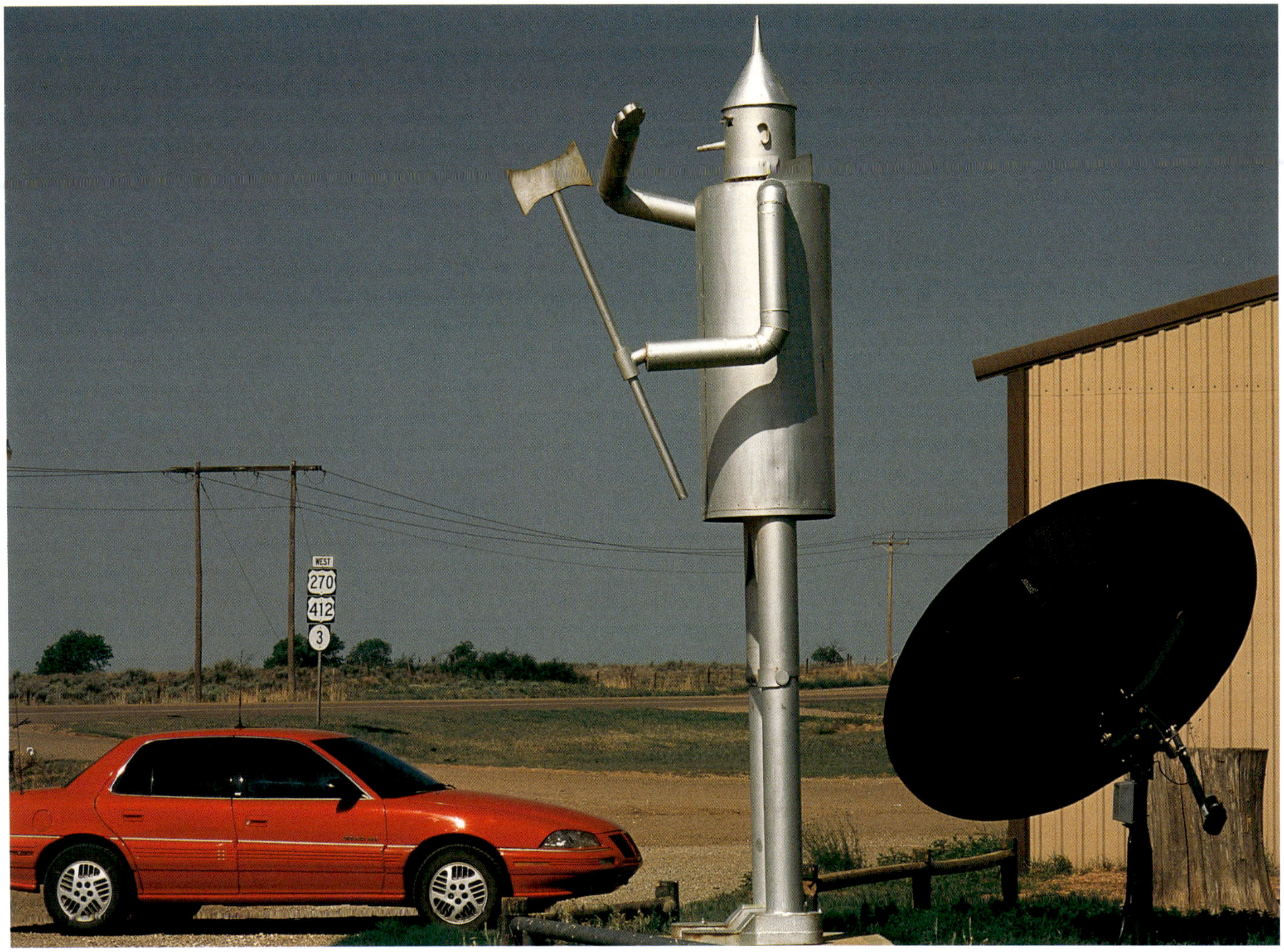

17-FOOT TIN WOODSMAN
Log Cabin Corner, Near Laverne, Harper County, Oklahoma

Plain. Shelterbelts protecting ranch houses from winter blizzards have disappeared, and the voluminous barns of the northern meridian have been replaced by sheds. Crystal City is also a northern extension of Hispanic culture. From here southward, people conduct their lives out of doors, Catholic churches mark the town squares, and images of the Virgin of Guadeloupe are seen on front lawns. Men embrace each other when greeting, a gesture unseen in the Dakotas.

With this changing climate and flora come different birds. In Kansas I saw the spectacular scissors-tailed flycatcher, its graceful tails twice the length of its body, in a memorable ballet of aerial courtship. Even more discernible changes occur fifty to one hundred miles north of Laredo where avian species from Mexico appear: the Crested Caracara, a curiously marked scavenger, the darkly elegant Harris' Hawk, and other exotic-looking birds such as the red-splashed Pyrrhuloxia.

Regardless of its geographical and cultural variability, the consistent, unifying force along the meridian is the struggle for existence at the margins of adequate rainfall. As weather influences land, land determines lifestyle. Residents of these prairie spaces have a relationship with the weather that is almost unfathomable to city people. Rainfall dominates their lives, determines their actions, and dictates economic survival.

In deference to the weather, my trips to the meridian were limited to spring and fall. Even so, one April I experienced a blizzard in the Sandhills, and in another April, in the

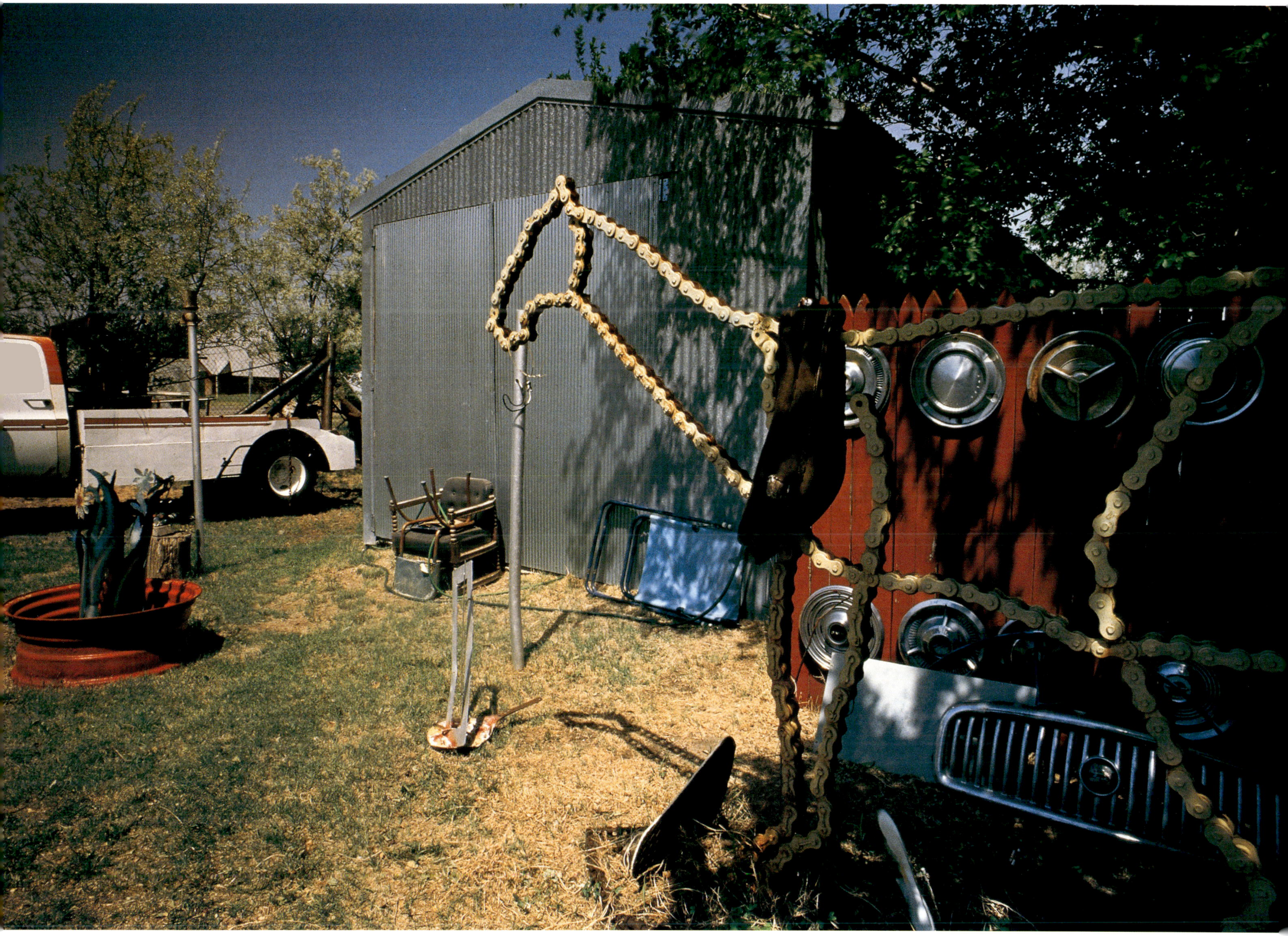

J. D. POWER'S HORSE
Gage, Ellis County, Oklahoma

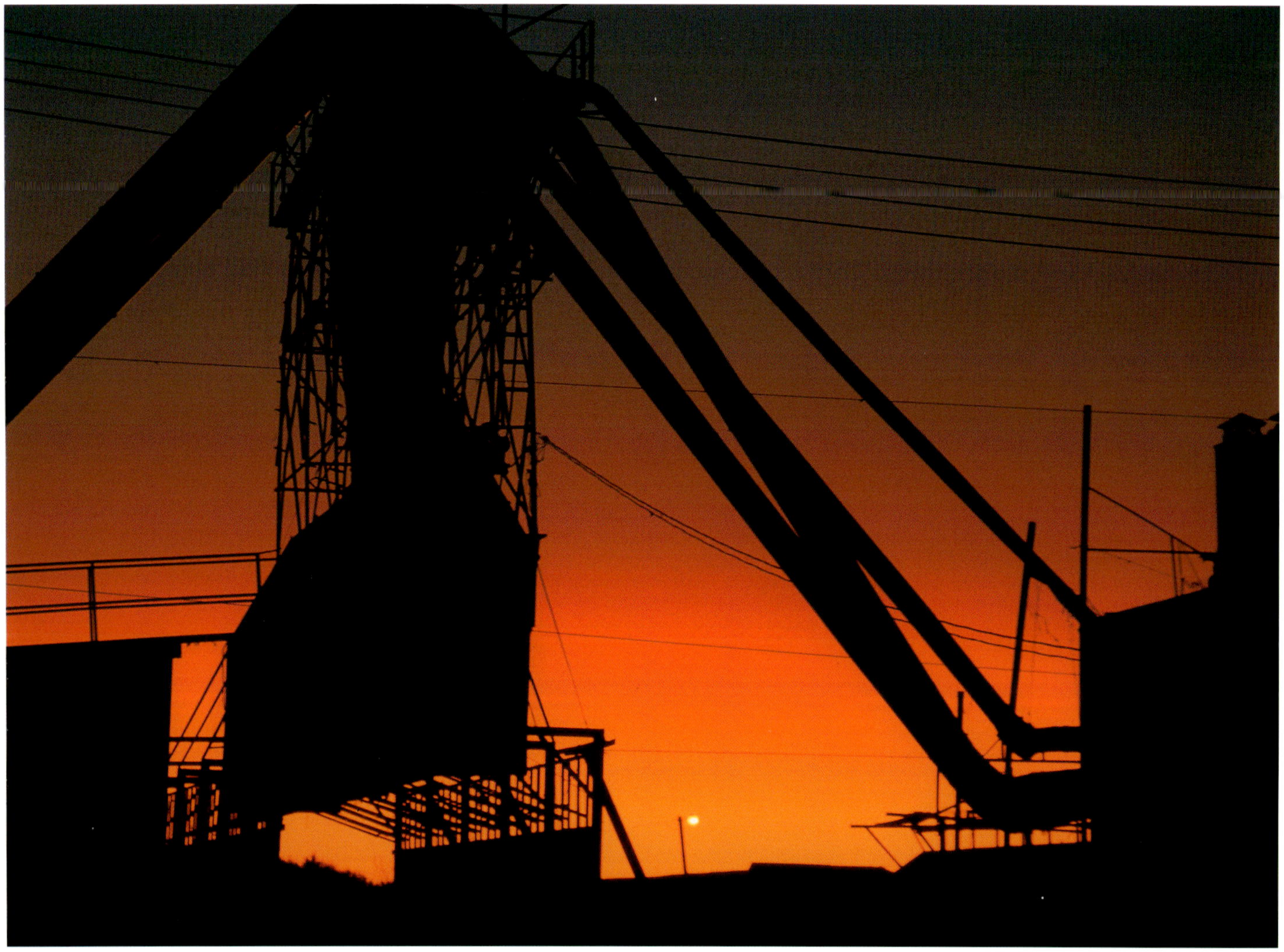

DAWN
Hollis, Harmon County, Oklahoma

Oklahoma panhandle, we had blast furnace weather: 106-degree temperature and 30-plus mile-per-hour winds. Subject to the volatile north-south shifting of the continental jet stream, this infamous weather has given rise to the saying, "If you don't like the weather here, just wait fifteen minutes."

Weather is the first and (virtually) only topic discussed by farmers and ranchers, as well as by shopkeepers, bankers, and townsfolk. In Nebraska I once overheard a mysterious discussion that went: "I got 38." A second said, "we had about 45," and a third added, "at our place there was only 28." Eventually deciphered, this conversation was about the previous night's moisture and the numbers were hundredths of an inch of precipitation. Farther south, in Texas, a pair of painted storefront windows read: "Welcome to Hamlin" and "Pray for Rain."

The 2,000-mile, empty chunk of America along the meridian seemed dusty, uninteresting, and brown most of the time. But sometimes, depending on the season and the rainfall, it was strikingly beautiful. Whatever its seasonal plumage, the loneliness combined with the silence of these spaces can be oppressive. Even today, that emptiness brings unease. To newcomers used to manicured, irrigated fields, the unease is sometimes exacerbated by the perception that large tracts are empty or "unused." But in reality, these seemingly unproductive fields are fallow, waiting a full year to collect sufficient moisture for the following season's wheat crop. Apparently vacant pasture might not be pasture at all, but a crop field banked in a government-sponsored soil conservation program.

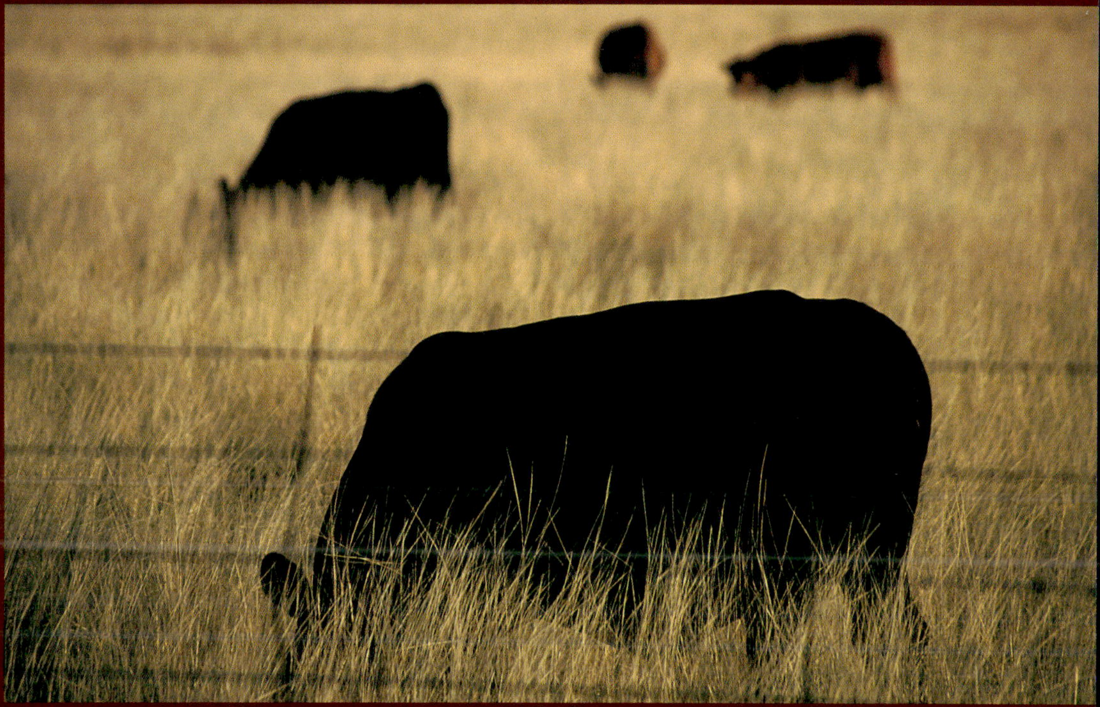

T E X A S

BLACK AND GOLD
South of Childress, Childress County, Texas

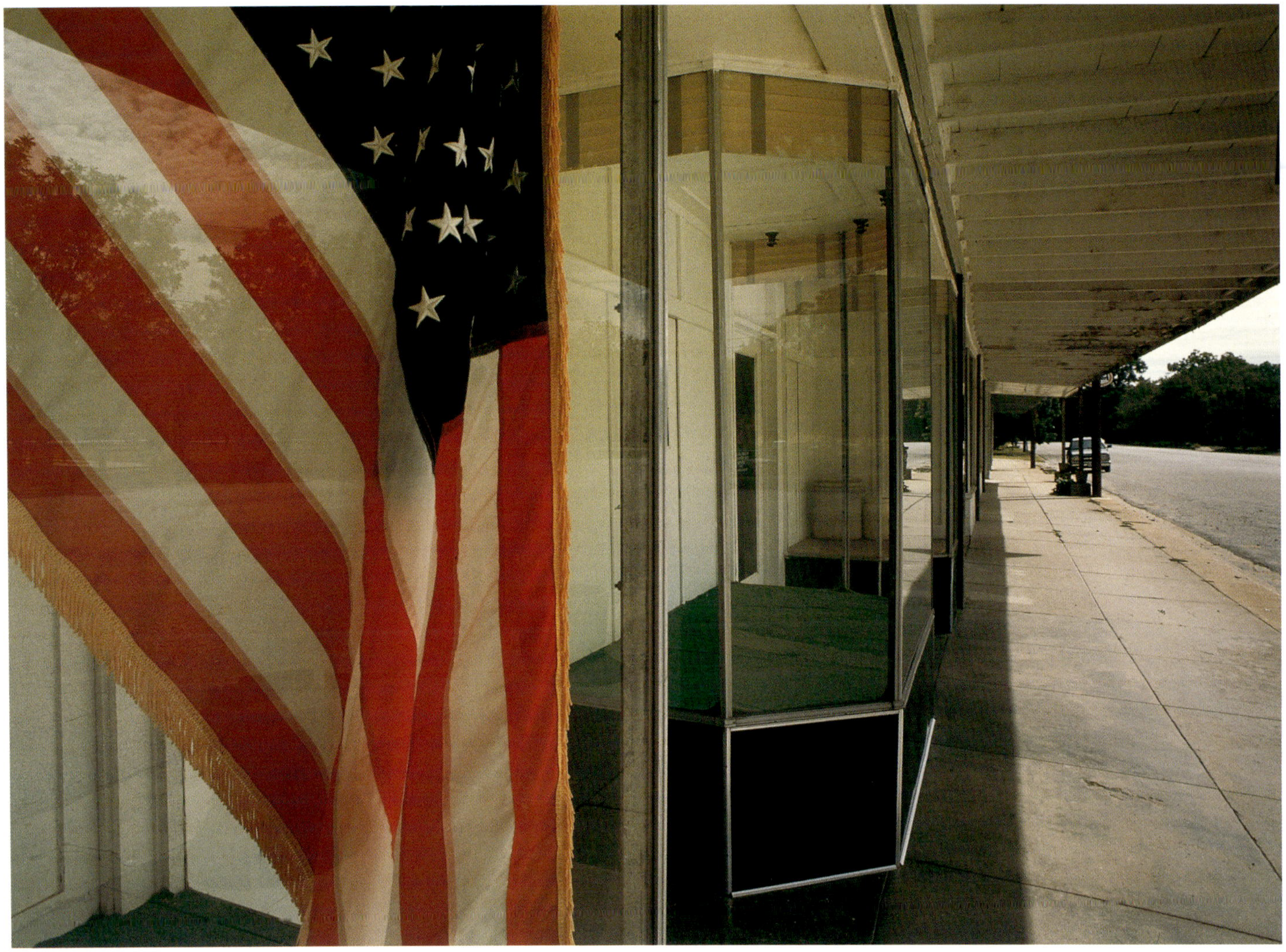

TELEGRAPH STORE
Telegraph, Kimble County, Texas

However desiccated, this land always harbors a bag of tricks, and in my wanderings I occasionally encountered incredible, albeit short-lived beauty — in the undulating breaks along major river courses, in Nebraska's Sandhills, and the Texas Hill Country, where increased moisture yields spectacular spring wildflower displays. On Highway 83 between Junction and Leakey the greenness of pine, manzanita, and mesquite was garlanded with miles of roadside ribbons of yellow wildflowers. Great stretches of gaillardia were punctuated by blue and pink clouds of salvia, milkweed, wild onion, and occasionally a bunch of fat-flowered penstemon.

Late in the year, rich grassland contours are softened with exquisite, velvety earth tones: gold, ochre, orange, and burnt umber. First snows are particularly lovely, brushing the yellow grasses into elegant, baroque visions of white-on-gold.

Whether rumpled and dusty, or ablaze with color, the land is unlike the East or the "wild" West, and its so-called emptiness is always filled with small miracles — the absolute blackness of a thunderstorm sky; bare tree branches abloom with a hundred yellow-headed blackbirds; the music of laughing coyotes dribbling across purple Texas twilight; and once, beyond a blind curve on a cold North Dakota afternoon, the quick, bedazzling brilliance of a thousand snow geese blanketing the swale like a winter snowfall come alive. These discoveries supplied me with visual adrenaline during the tedious miles of driving that separated photographic moments.

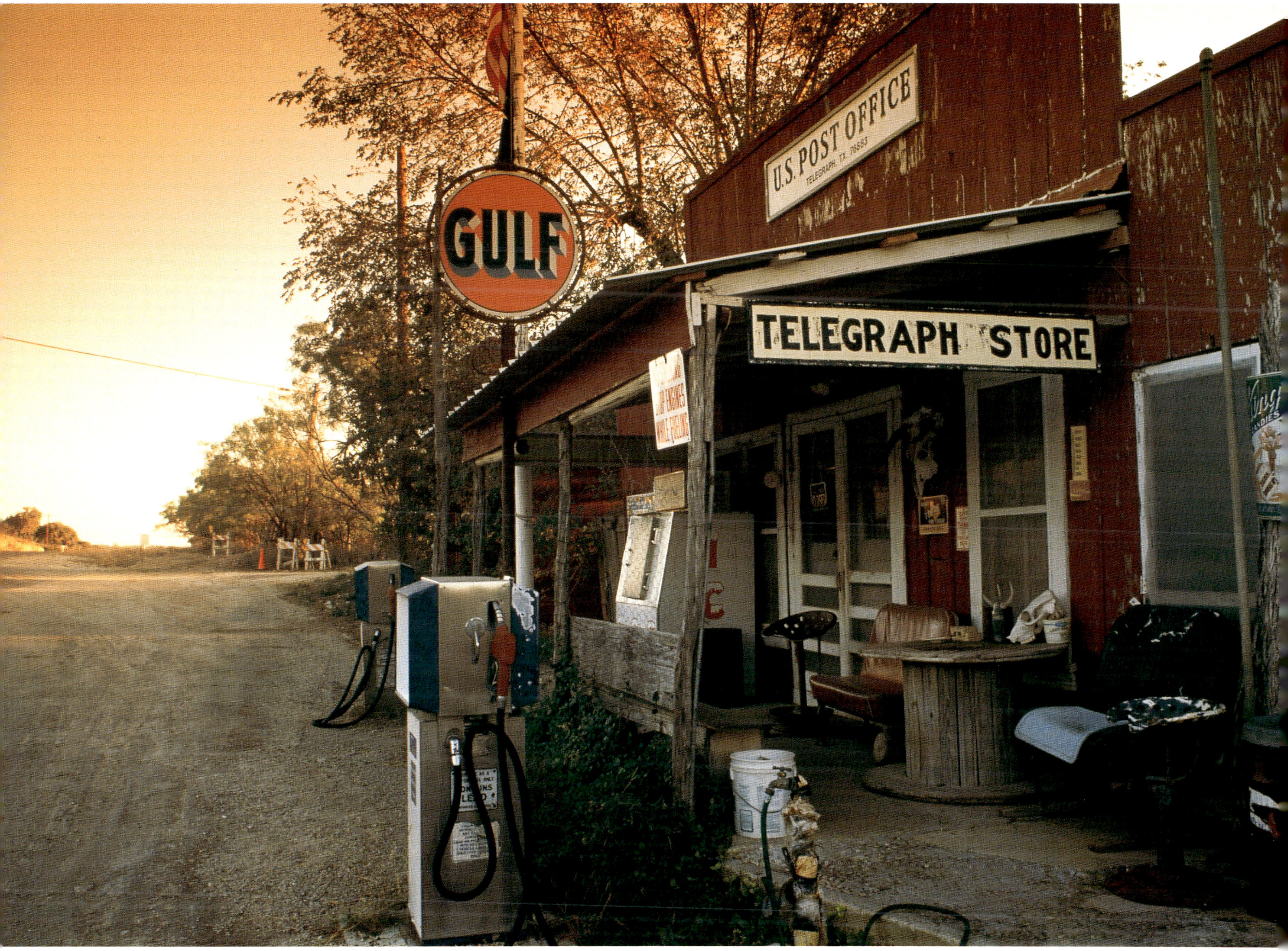

USDA OFFICE
Menard, Menard County, Texas

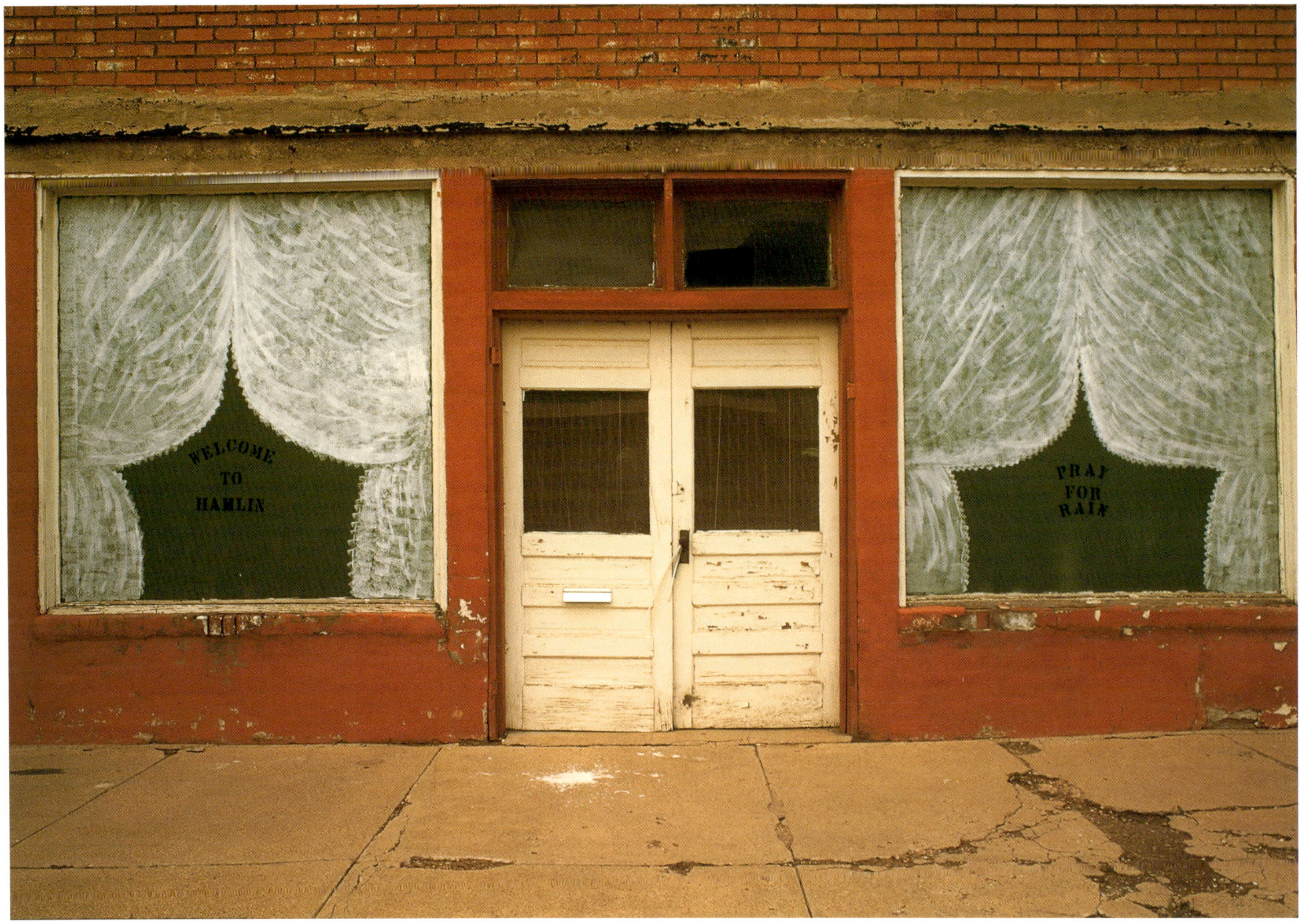

PRAY FOR RAIN
Hamlin, Jones County, Texas

Crossings and Endings

On Highway 152 west of Sweetwater, Oklahoma, I found the "100th Meridian Stop" gas station so precisely on the 100th meridian, that its convenience store was in Oklahoma and its gas pumps in Texas. After I explained my project, the owners gifted me with a "100th Meridian Stop" cap and coffee mug. Charlie Hagerman walked me out south and easily found one of the markers delineating the 136-mile portion of the meridian that borders Oklahoma and Texas. With only two inches of rain in the previous seven months, the pylon with its bronze disc was clearly visible above the drought-stunted wheat. Emplaced at one-mile intervals by a 1932 decision of the U.S. Supreme Court to end years of bickering between local residents of Oklahoma and Texas over the location of the border, this is the only portion of the 100th meridian that is marked. Charlie then told me the story of finding one of his prize bulls stone dead, unmarked and still standing after an electrical storm, only its nose blown off by a lightning strike. Dramatically gesturing at a spot between our feet just a stone's throw from the 100th meridian, he said, "Right here! That bull cost me thirteen hundred dollars and I'll *never* forget where it was killed!"

Although cattle are raised all along the meridian, central Kansas is serious meat country. Just east of Dodge City, on Highway 50, there is a scenic overlook advertised to motorists coming from either direction. At the overlook a surprise — not a picturesque green valley or a postcard-perfect river

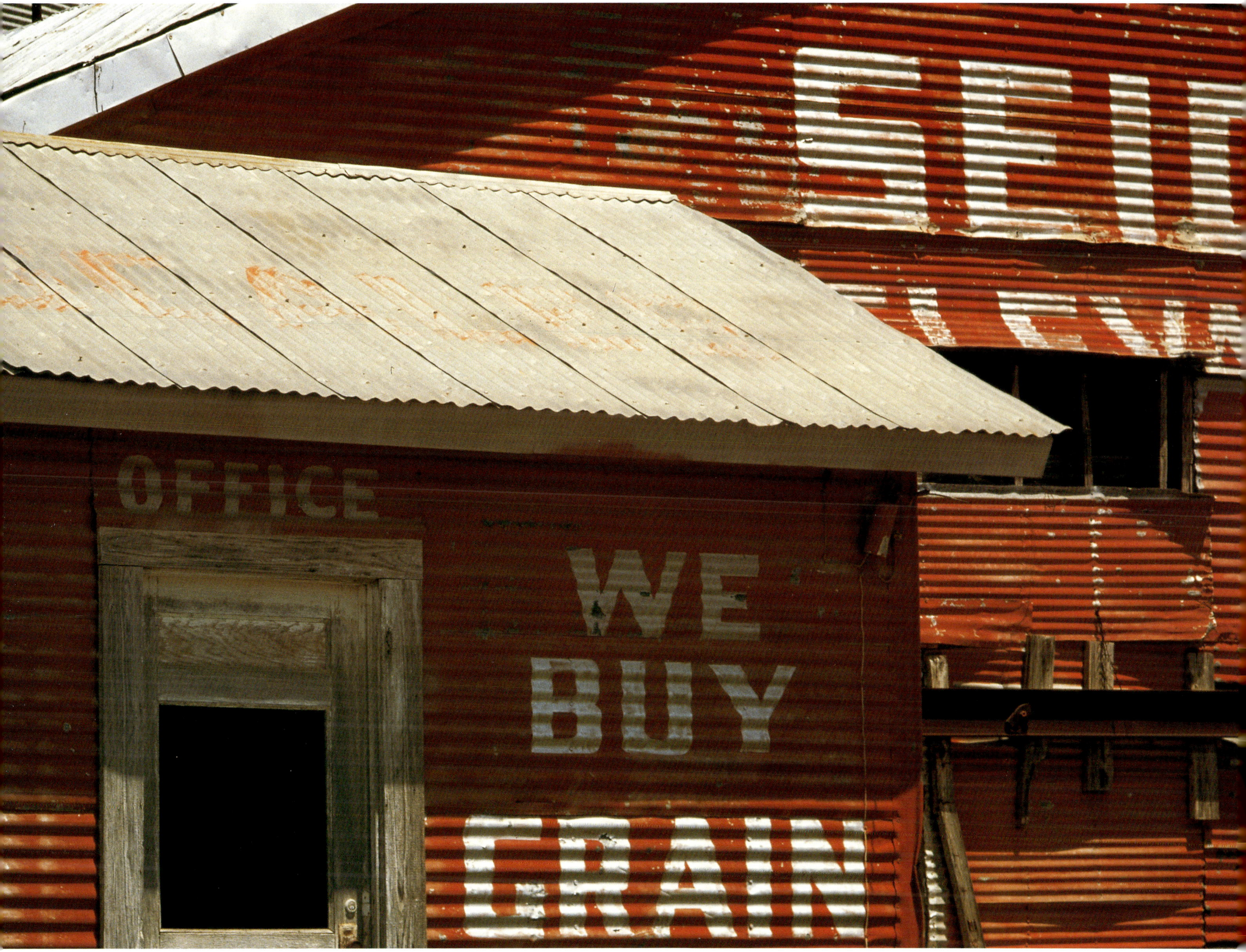

"WE BUY GRAIN"
Ballinger, Runnels County, Texas

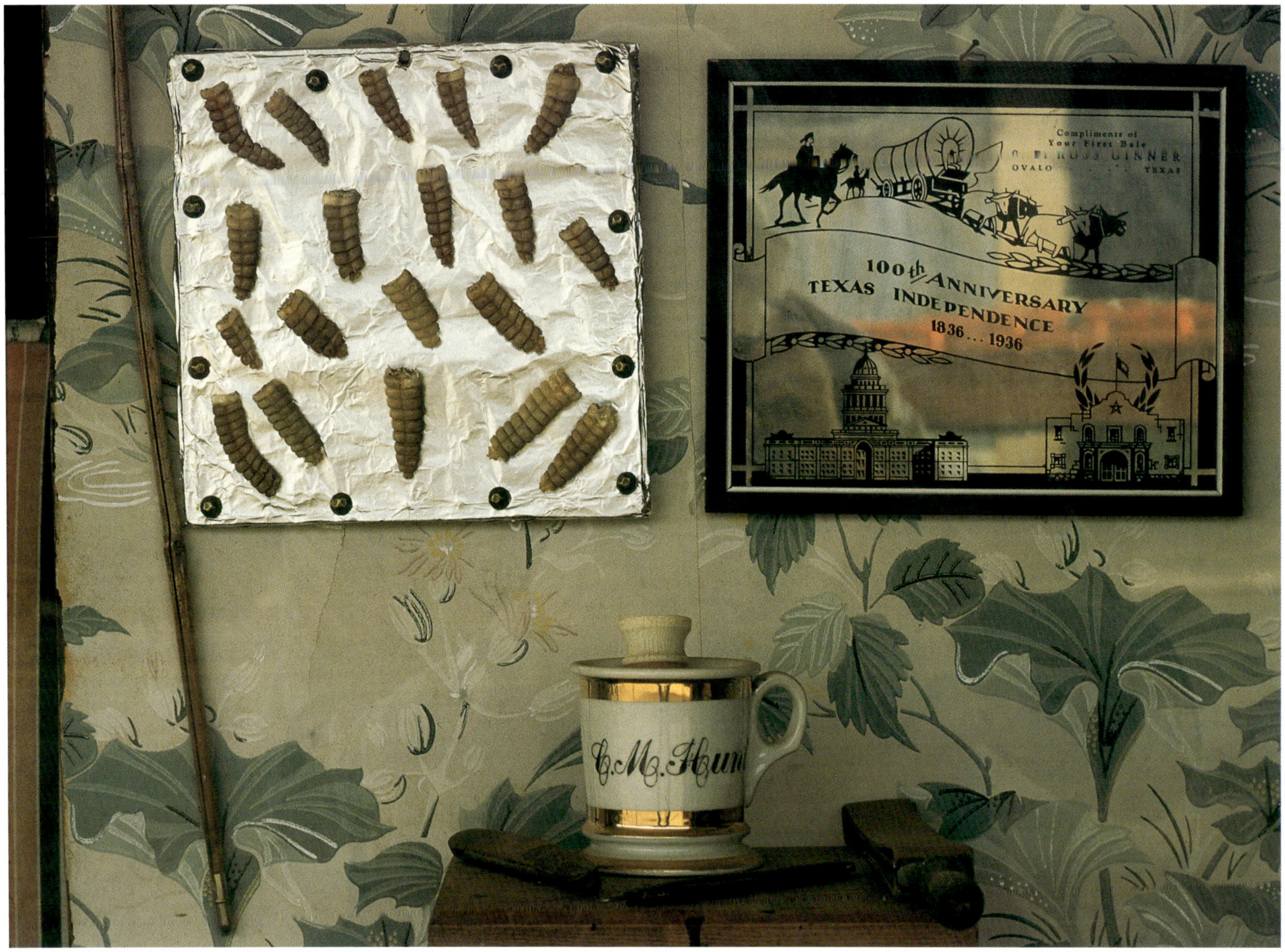

100TH ANNIVERSARY
Bradshaw, Taylor County, Texas

curving toward a distant horizon, but smelly stockyards — cattle pens, as far as the eye can see. The sign's statistics were impressive: 390,000 head pass though the auction blocks each year. Every day, 750 truckloads of cattle roll into this ultimate cowtown, and an equivalent number of refrigerator trucks leave filled with packaged meat. It helped to explain the billboard on the other side of Dodge, the one that urgently proclaimed, "Eat BEEF. The West wasn't won on salad!"

Dodge City is, by accident, almost right on the line, but the town of Cozad, Nebraska, is probably the only town deliberately platted exactly on the 100th meridian. In 1882, professional gambler and promoter John Cozad thought that this unique location would capture the imagination of potential land buyers. Although he successfully launched the town, two years after its founding, Cozad was forced to leave hurriedly after shooting a local rancher in a dispute. He was never seen in Nebraska again, and his wife and two young sons disappeared soon after. Although Cozad senior was eventually exonerated, the whereabouts of the Cozad family after fleeing Nebraska were a mystery until 1956.

Under a new identity, Robert Henri, the younger Cozad son became an influential American painter whose life was the subject of Mari Sandoz's *Son of a Gambling Man*. Also using an alias, Cozad senior continued his land development schemes until he died, his new identity still undiscovered. The hotel he built is now a museum telling the intertwined stories of his painter son and the community he founded where the railroad crossed the 100th meridian.

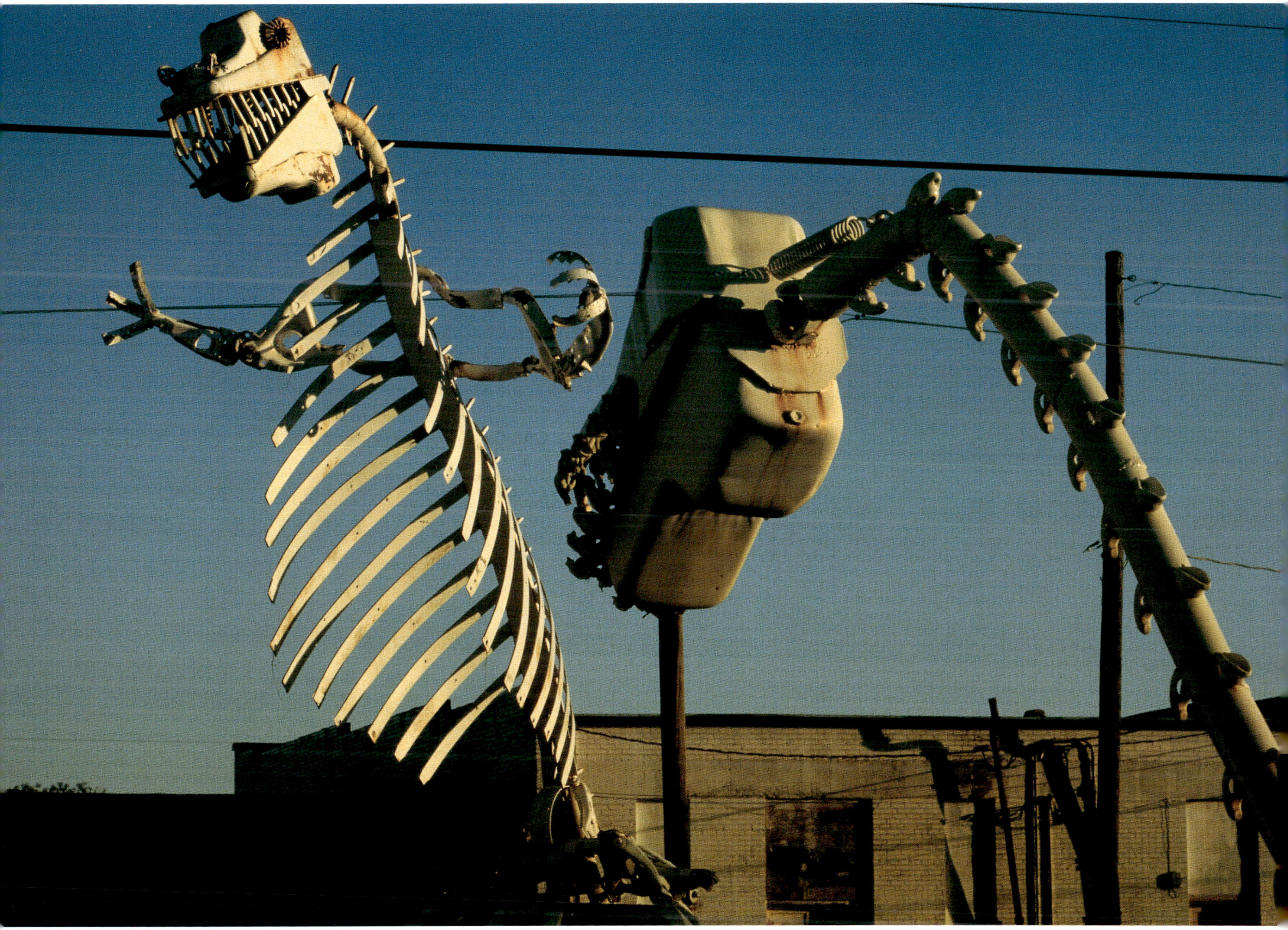

TYRANNOSAURUS REX
Stamford, Jones County, Texas

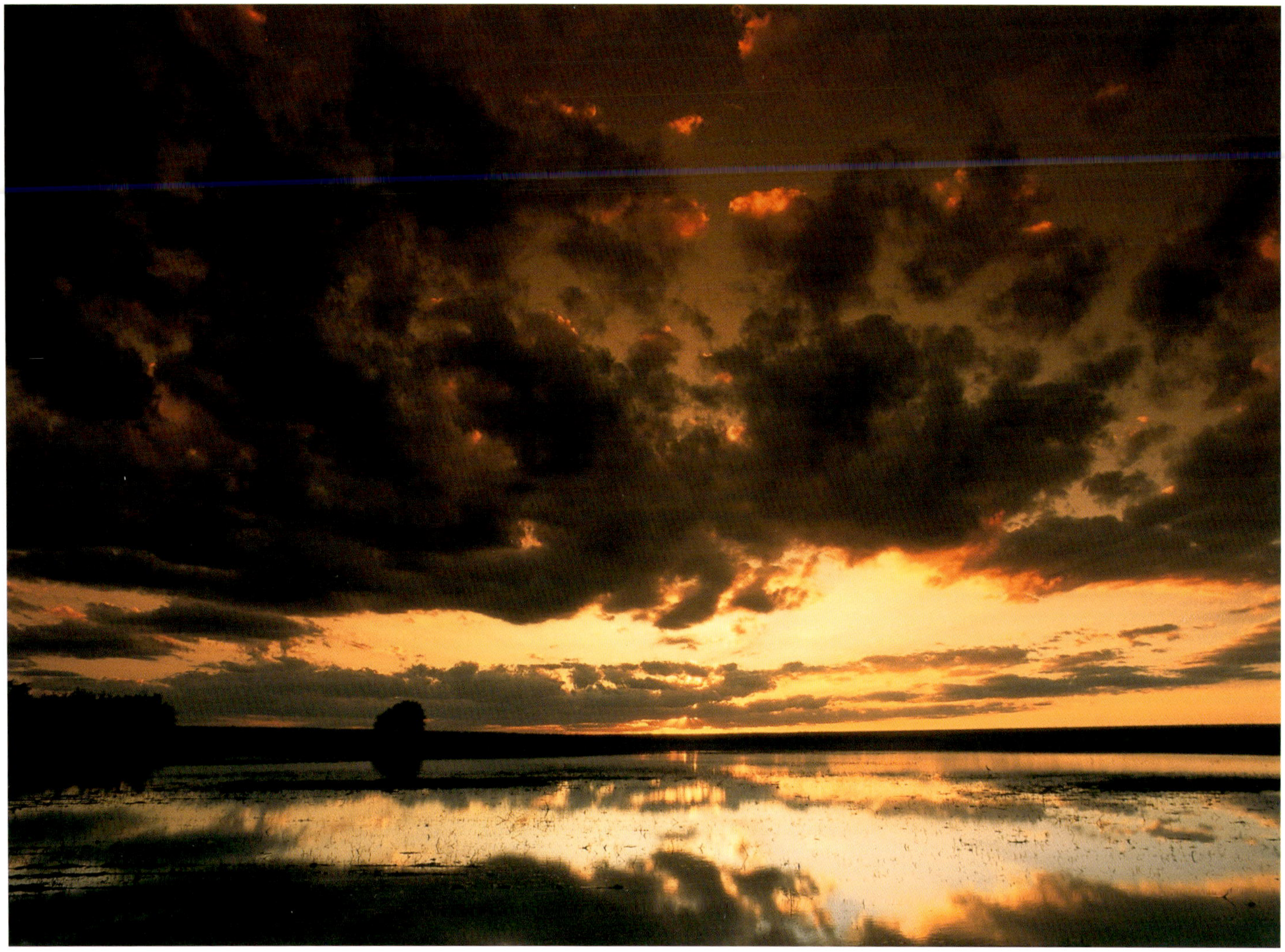

FLOODED FIELD
Noodle, Jones County, Texas

At the top of the U.S., the 100th meridian crosses the Canadian border in Rolette County, North Dakota, about two miles east of the International Peace Garden on the property of a man named Glen Honsey. I found Glen's mailbox, probably the northernmost on the American meridian, emerging from the roadside among thick grass, yellow-blooming jack-in-the-pulpit, and wild roses. Honsey wasn't home, so I had to search for the crossing point. Starting at the Highway 281 border station, I found the cleared strip marking the border by driving a convoluted mix of gravel roads and traversing a fenced pasture. Walking the border, I located a silver pylon stamped "Marker No. 913" by the U.S. Border Commission, and after carefully pacing off 744.2 feet west from the marker pylon, I reached the exact point of the meridian crossing. There was ordinary pasture on the U.S. side and a tangled alder thicket on the Canadian side. Without a marker the spot wasn't interesting enough to photograph.

At its southern end, the meridian border crossing is about fifty-five miles northwest of Laredo in Webb County, Texas, where the Rio Grande flows sharply northwest-southeast, bordering the property of Robert Briscoe, a former governor of Texas. The crossing point cannot be seen from the road. Much of this part of Texas, including Briscoe's ranch, is fenced with formidable, ten-to-twelve-foot high, multi-strand barbed wire, which usually encloses private hunting lands stocked with exotic game. Texans take these fences seriously, and they are earnestly patrolled by air and on the ground, one of

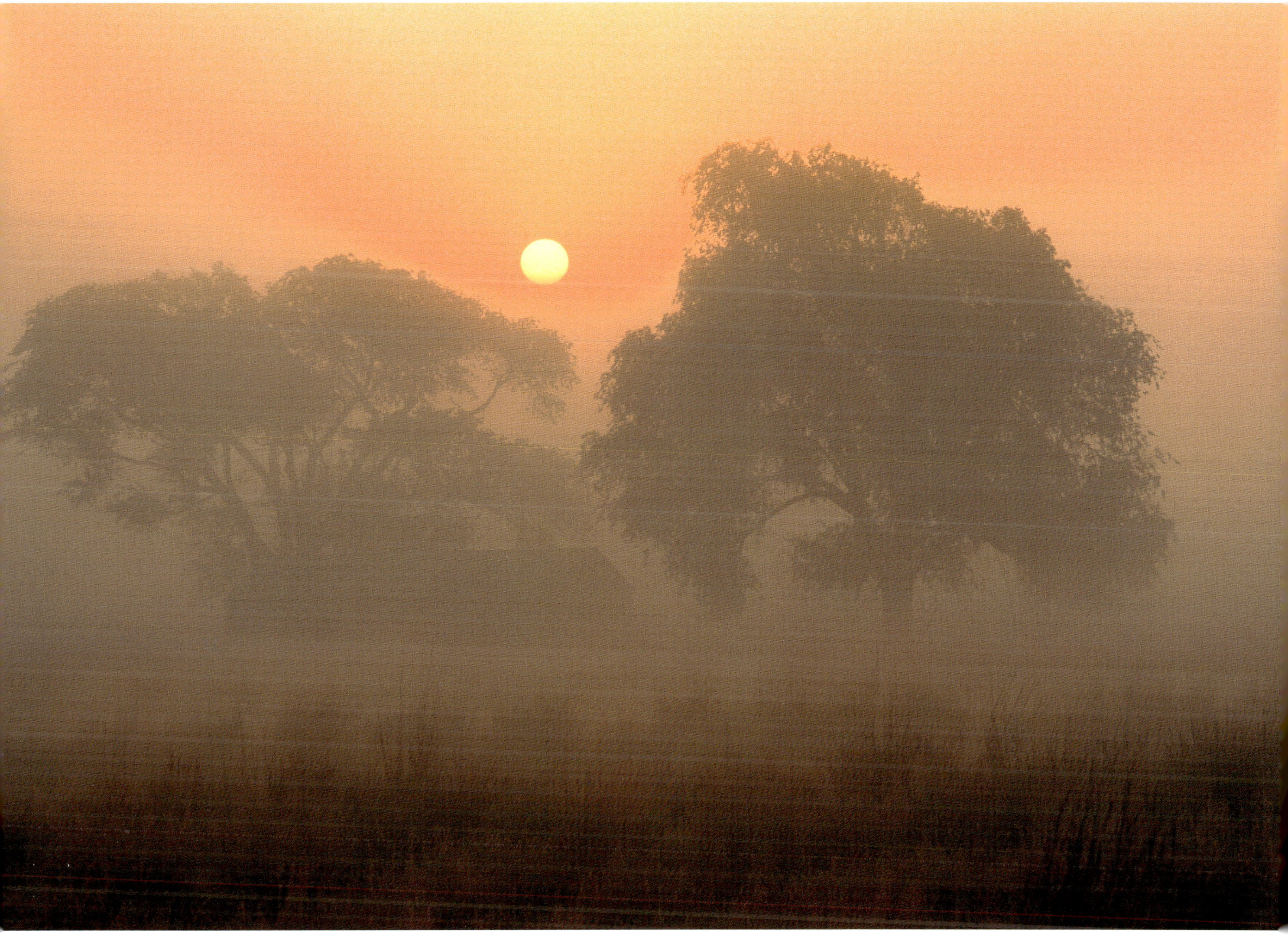

MISTY SUNRISE

West of Childress, Childress County, Texas

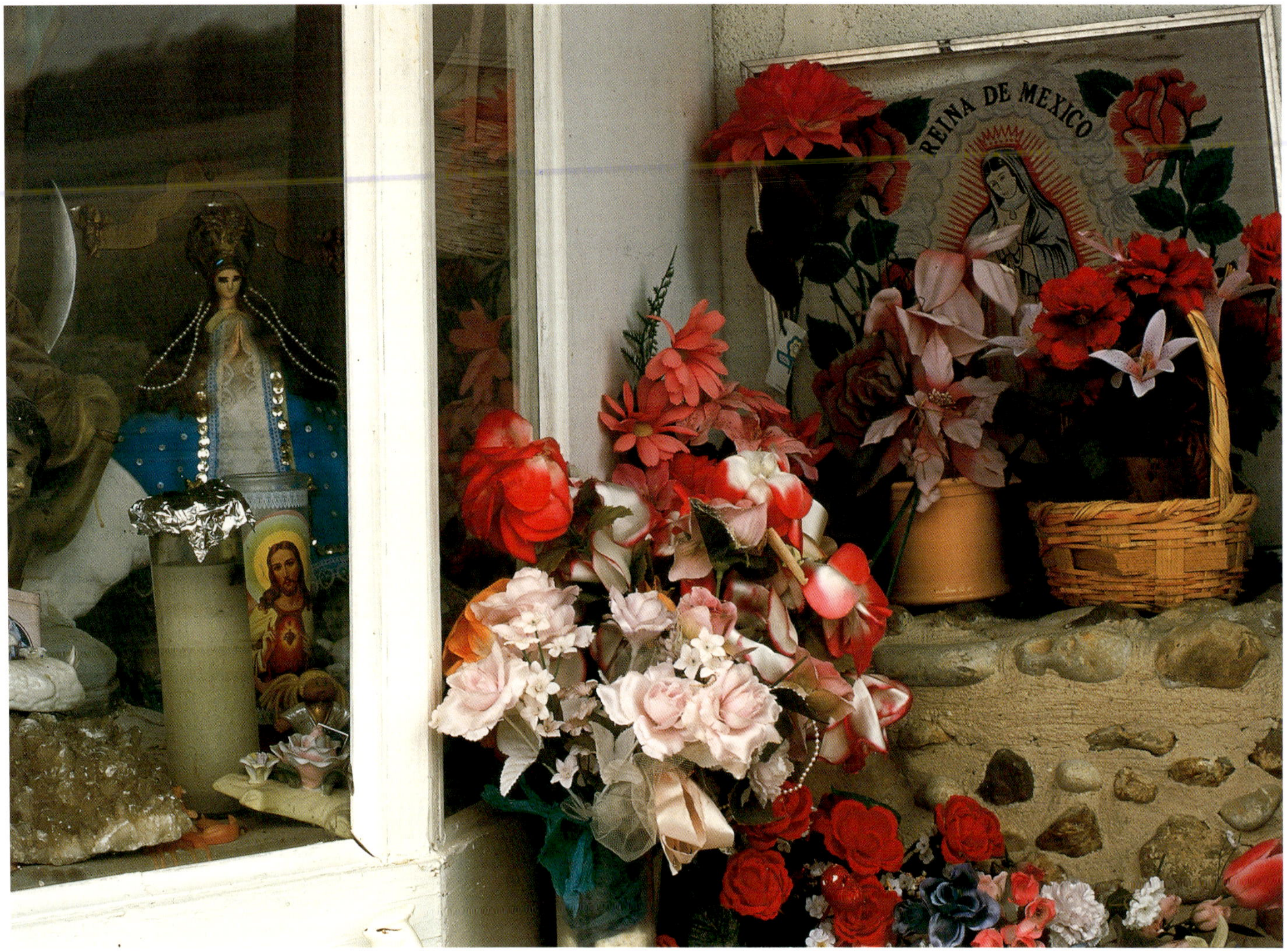

ROADSIDE SHRINE
Uvalde, Uvalde County, Texas

the less friendly aspects of the state. I never bothered to contact Briscoe to gain entrance.

The Passage of Time

In the nineteenth century, farmers and ranchers were dependent on towns for storage and distribution of their products to the rest of the nation. Over the years, however, these towns began to disappear. Cheap transportation and paved roads destroyed the isolation created by distance. A town thirty miles away by automobile became more accessible than one six miles distant had been by horse and wagon. Grain and cattle were consolidated at fewer, more distant locations, and the need for manpower was drastically reduced by the ever-increasing mechanization of farming. Populations shrank dramatically, and many towns became obsolete or "culturally extinct," as they are sometimes described. During a chance encounter, I received a touchingly personal recitation on this topic.

What remained of Truscott, Texas, in 1993 was a single east-west street whose northern edge was a two-block stretch of collapsing storefronts in a sea of weeds. It was hard to tell exactly where the wreckage ended and the weeds began. Farther east were the solemn stone-on-stone remains of a bank. Opposite the storefronts, a decrepit backstop was emerging from an overgrown field or perhaps returning to it; the weeds were so high I wasn't sure.

I was photographing the backstop when Tommy Tapp approached, looking for all the world like the town's only citizen, since there was no one else there. Tommy volunteered that in 1924 he'd been brought to

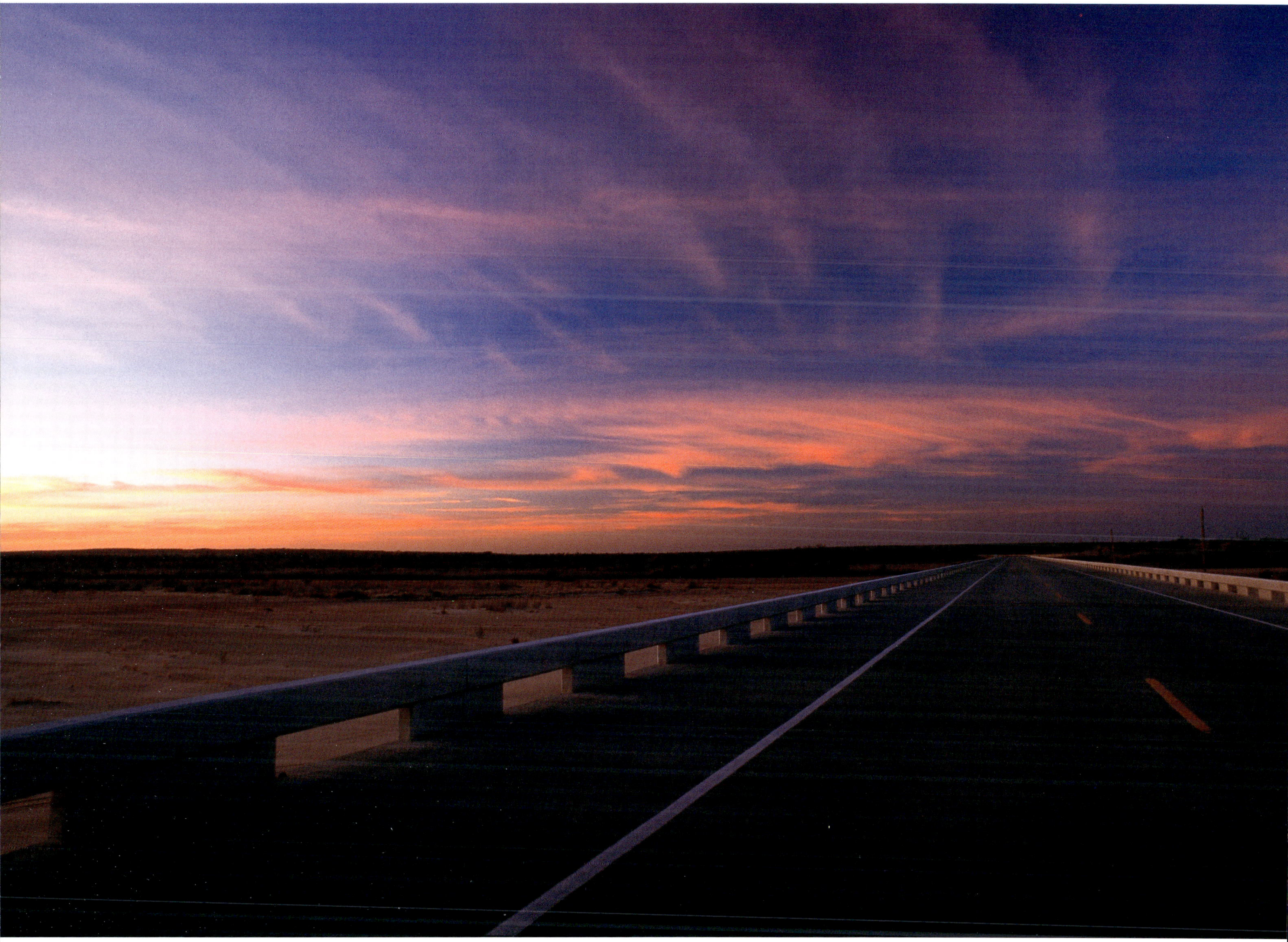

HIGHWAY 83 CROSSING
THE RED RIVER
Childress County, Texas

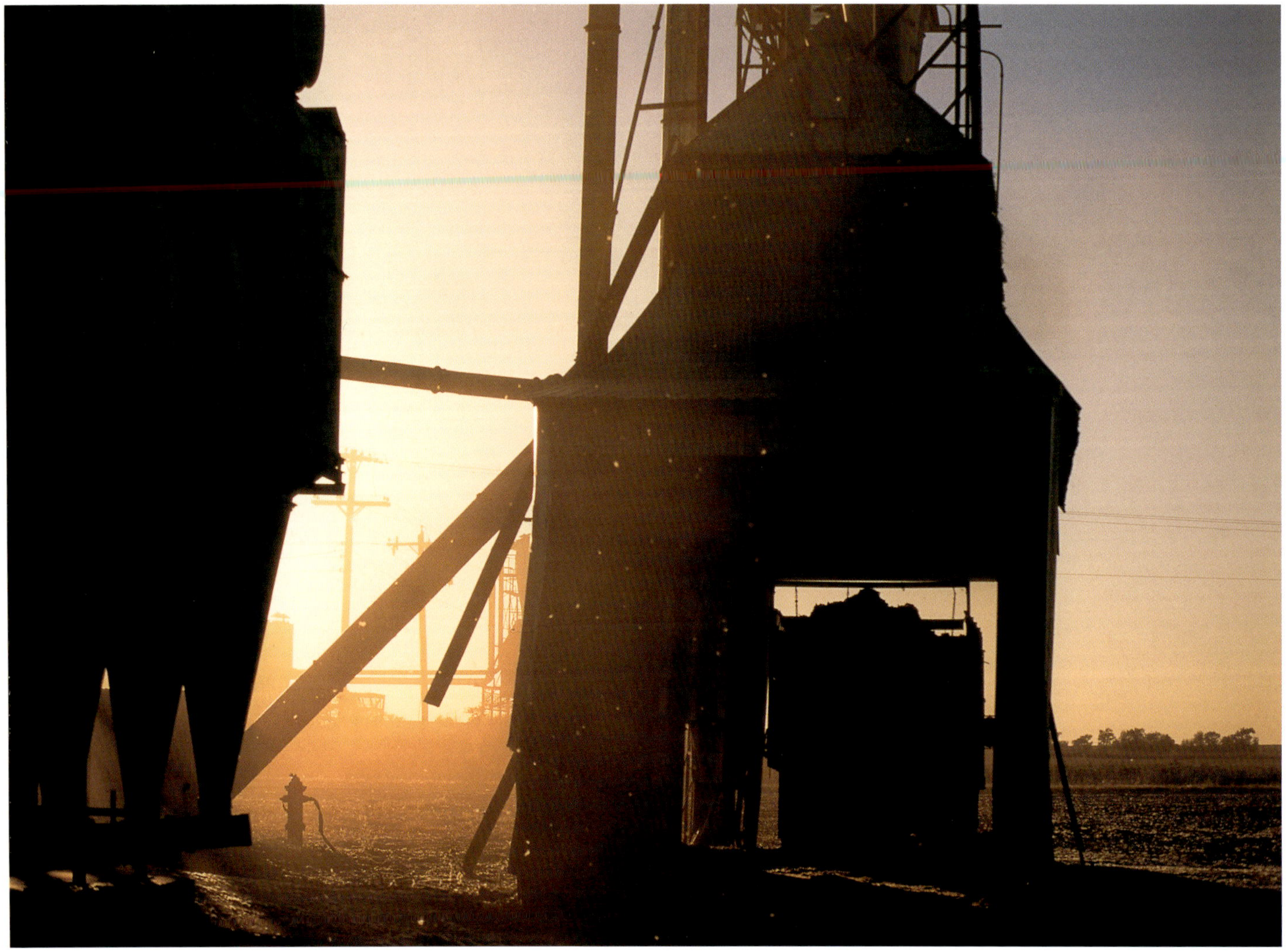

UNLOADING RAW COTTON
Location Unknown, Texas

Truscott from the country to attend school, at age ten. Before that, his life was hoeing and picking cotton. Pointing imprecisely to the piles of weathered wreckage, he continued, "There's the old cleaners, and next door was the barber shop where cowboys used to come and bathe, getting their clothes cleaned up for Saturday night. Town used to have four-hundred kids, now even the school is gone. Used to be, Henderson Ranch would ship cattle from here by the trainload . . . wheat and cotton too . . . the decline started in the thirties; in the forties, people moved away to work the war plants. After the war, machines came and replaced people hoeing and picking cotton and shucking corn. Used to be good people working together. Now, you can't get three people together to talk, let alone agree on anything."

Healthy towns flourish at the expense of the dying. As commercial functions coalesce at more centralized locations, schools and social services also tend to consolidate at those locations, further strengthening the survivors, which usually had some fragile economic or geographic advantage — a key highway intersection, a hospital, or perhaps a county seat. In Oklahoma I was tutored on small town economic realities with a painful anecdote about a town's economy that almost collapsed when the local hospital's chief surgeon transferred to a competing hospital in a neighboring county. After hiring away three doctors from his former facility, this surgeon almost shut down the hospital, which was central to the town economy. Only the townspeople's Herculean effort to hire replacement doctors saved the

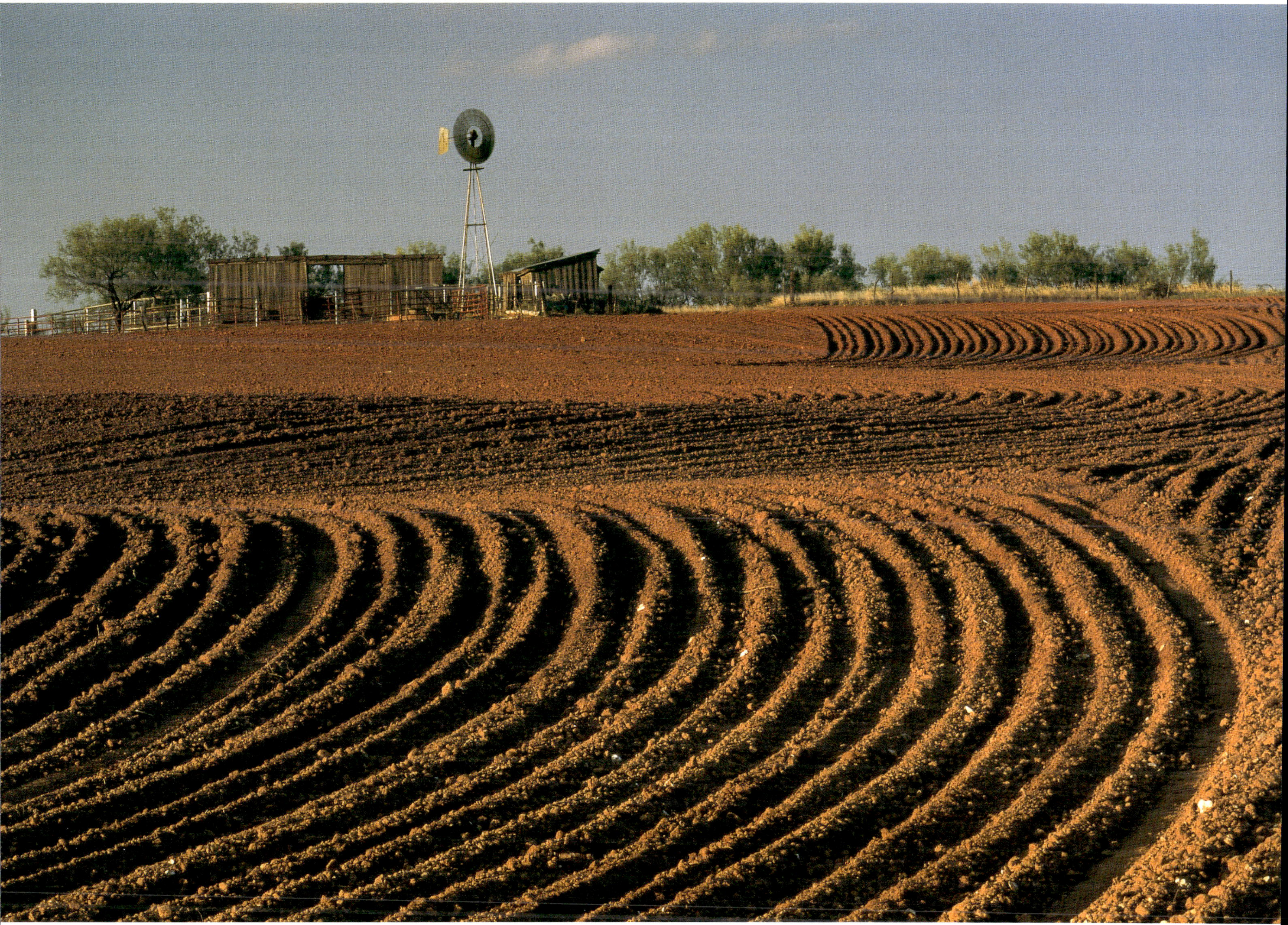

FRESHLY PLOUGHED
North of Paducah, Cottle County, Texas

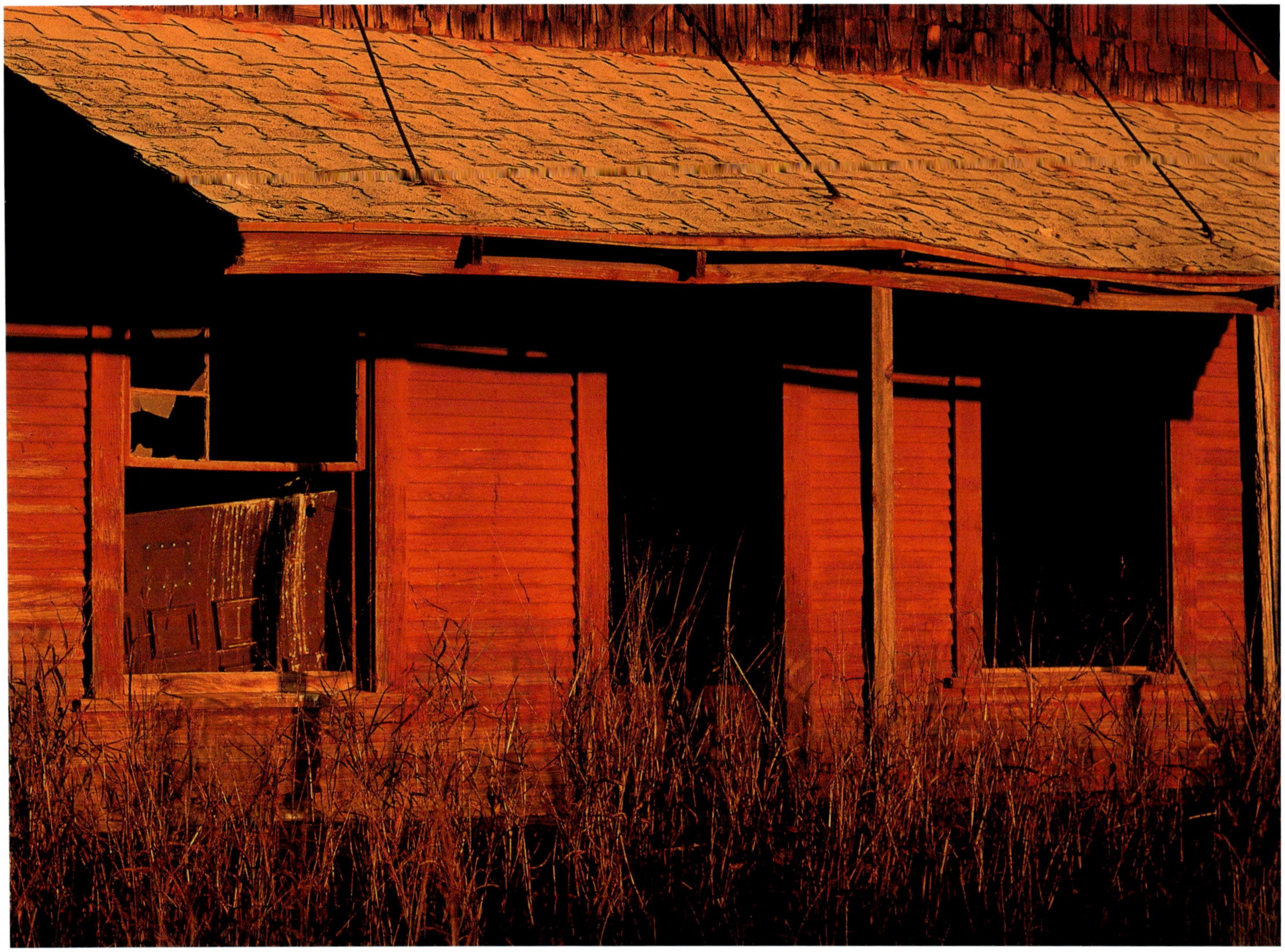

RED HOUSE
Cottle County, Texas

day. Young MDs and their spouses are not easily attracted to small town practices in rural Oklahoma.

Unlucky towns die slowly, populated at the end by isolated enclaves of old people, descendants of earlier generations, who lack the finances, energy, or emotional fortitude to start new lives during their seventh or eighth decade of life. Since unchecked population flight results in fiscal distress and physical neglect, the survivors often live in townscapes littered with the legacy of pioneer thrift: the rusting residue of appliances, damaged trailer houses, flatbed trucks, trailers, derelict cars, pickup trucks, wrecked machinery with pulleys dripping filaments of weathered belts, and rusted fifty-five-gallon drums overflowing with all manner of scrap metal, bolts, nuts, and small parts.

The dusty store-front museum ubiquitous to every small town is the usual repository for some of this old "stuff," and at first glance the heartland seems to show a concern for the past that borders on institutionalized sentimentality. But after spending time on the meridian, one realizes that preoccupation with the past has less to do with nostalgia than with the awesome human drama enacted there in the nineteenth century, one that is still playing out in surprising ways in unlikely places.

Lost on a back country road searching for a bridge across the Red River into Texas, I stumbled upon a family of fourth generation ranchers adjusting to new realities at "Bar-S Monogram Industries." Smack in the middle of remote ranch country, this ten-year-old business used computerized machines to custom silk-screen and

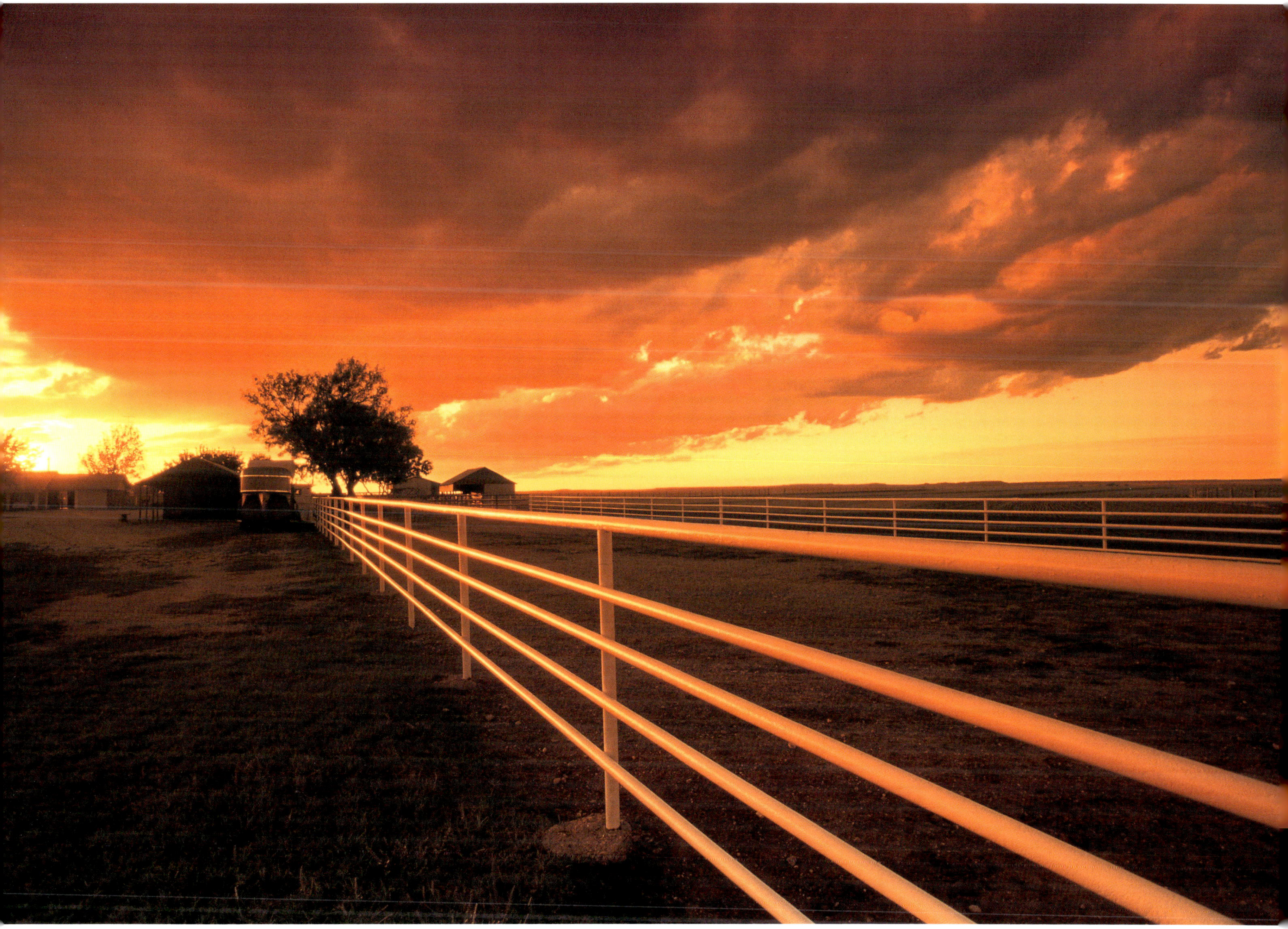

SUNSET PERSPECTIVE
Old Glory, Stonewall County, Texas

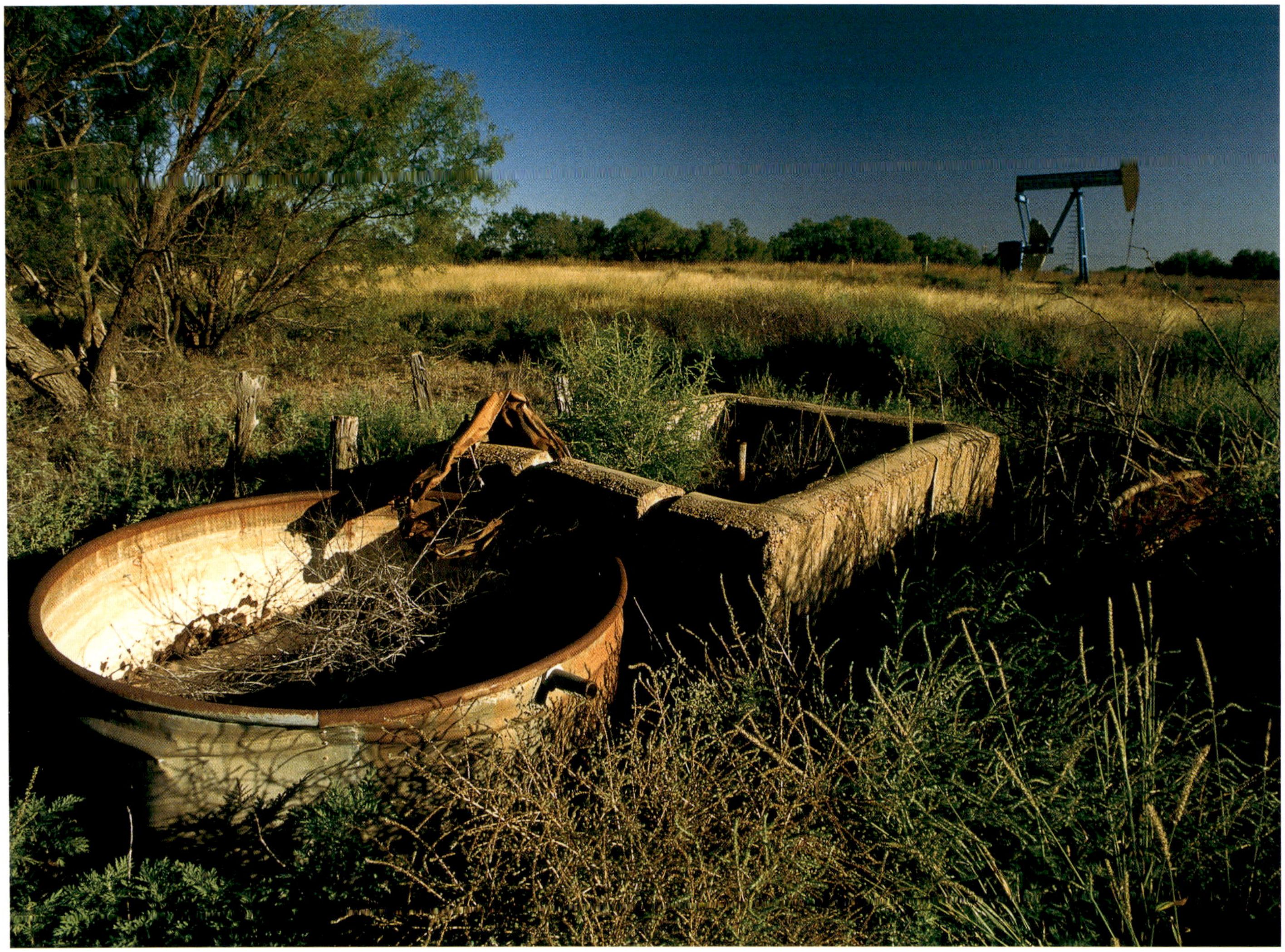

TEXAS LANDSCAPE
Sagerton, Haskell County, Texas

embroider T-shirts, advertising giveaways and athletic uniforms for schools, service clubs, and businesses. The flexibility shown in establishing this business is the more remarkable given the family's hundred-year history of excellence in Hereford production, proudly demonstrated by a floor-to-ceiling display of awards for breeding prize cattle. The office wall was witness to the fact that in today's world, success in the cattle business is dependent on factors other than competence and hard work. The juxtaposition of cattle breeding and computers was an interesting display of adaptability.

People and Stories

I liked the meridian people. I found them to be generally quiet and respectful, mindful of their own business, not overly talkative, curious about but not necessarily welcoming of strangers. Born of generations heavily focused on persistence and endurance, they seem emotionally guarded and averse to frivolity and gaiety, yet possess an ingrained sense of humor honed as a survival tool. They are skillful and mechanically inventive, with habits of hard work, honesty, thrift, and humility, learned early and well in a land that tolerates no laziness and little generosity.

Before starting the project I was apprehensive about being welcomed as an outsider. My experiences elsewhere told me that strangers in rural areas are generally viewed with suspicion. Farmer friends insist that this seeming paranoia is well-founded: prowling strangers are known to "liberate" dusty "antiques" from unguarded barns

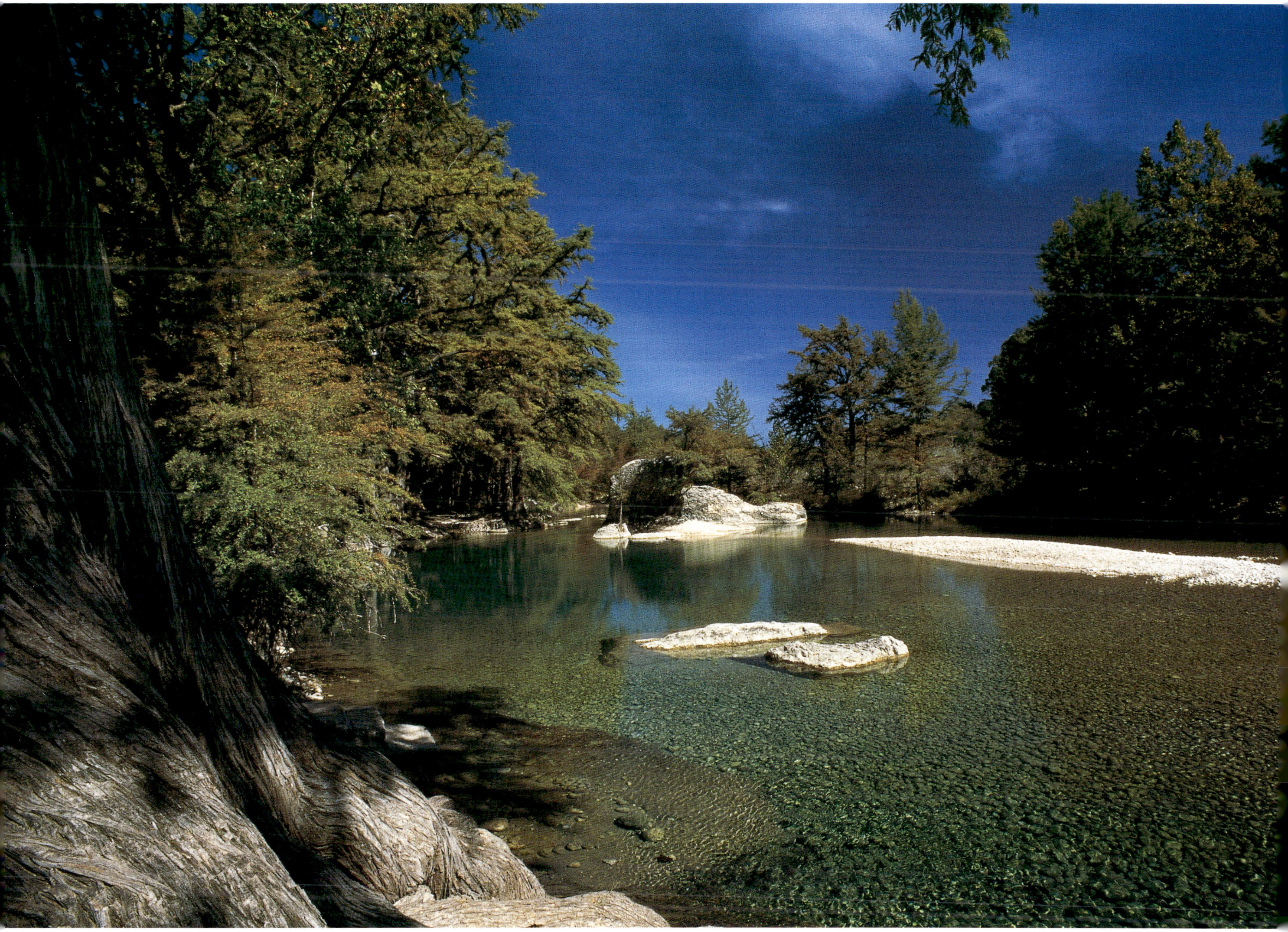

RIO FRIO
Real County, Texas

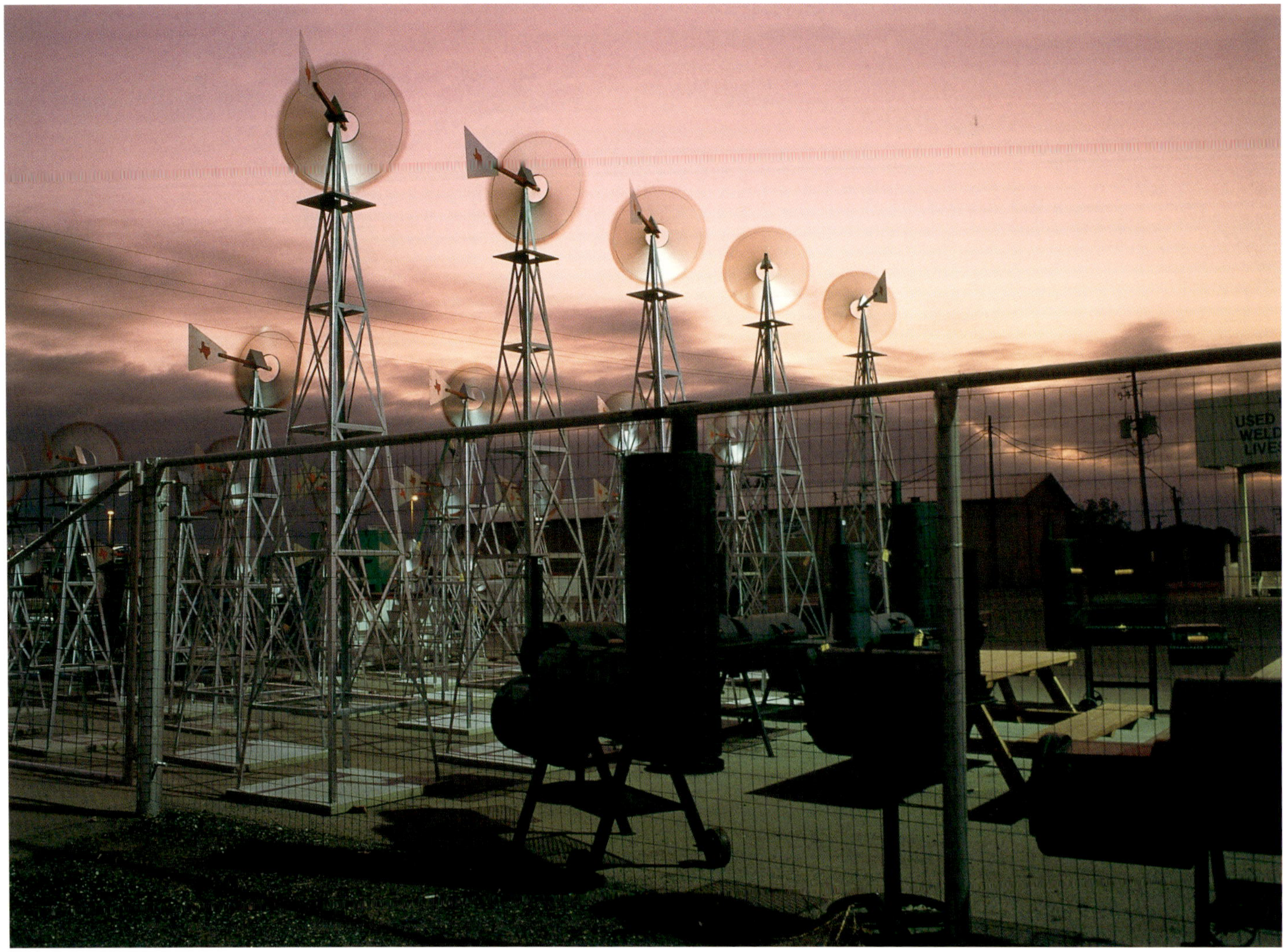

WINDMILLS AND BARBEQUES
Childress, Childress County, Texas

under the assumption that remoteness justifies behavior commonly called theft in urban areas. Ignorant of ranch protocol, city people are often arrogant in their failure to ask permission to picnic, camp, or hunt on ranch land, and are notorious for leaving debris, smoldering cigarette butts, and opened gates behind them. For these reasons, I was concerned about hostility from local residents.

California plates and the absence of mud on my truck clearly marked me a stranger, so I was surprised at the ease with which my presence was greeted. While curiosity seemed restrained ("What are you doing here?" came out as "My, you sure are a long way from home!"), cordiality increased as I explained my project and blossomed into open warmth at the notion that someone from far away could be interested in their lives.

Once, searching a cold South Dakota dawn for the foreground to a perfect sunrise, I passed a place with barely discernible but interesting silhouettes. It was unsociably early and very dark when my reluctance to request permission to enter was interrupted by the headlights of a departing school bus. "Aha! Somebody is awake." Putting my trepidation aside, I knocked. An unshaven rancher, barely awake, appeared in the expanding wedge of yellow light. He was half-dressed, one hand clutching unbuckled trousers at his waist. "What can I do for you?" When I said I was photographing a picture story on the area and asked if it would be okay to enter the property to shoot, he answered, "Fine," and closed the door.

His laconic answer typified the way I

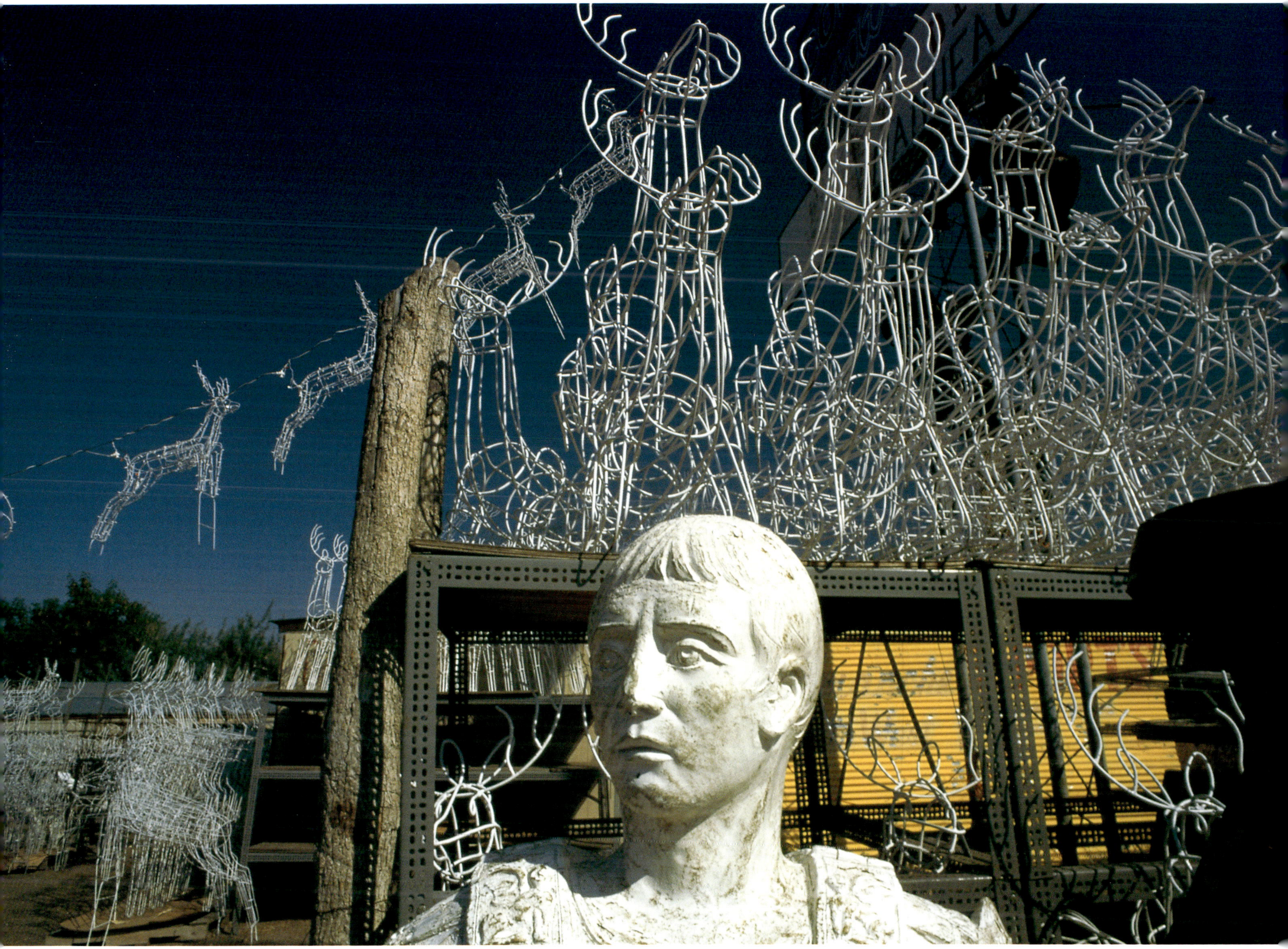

CHRISTMAS CAESAR
Laredo, Webb County, Texas

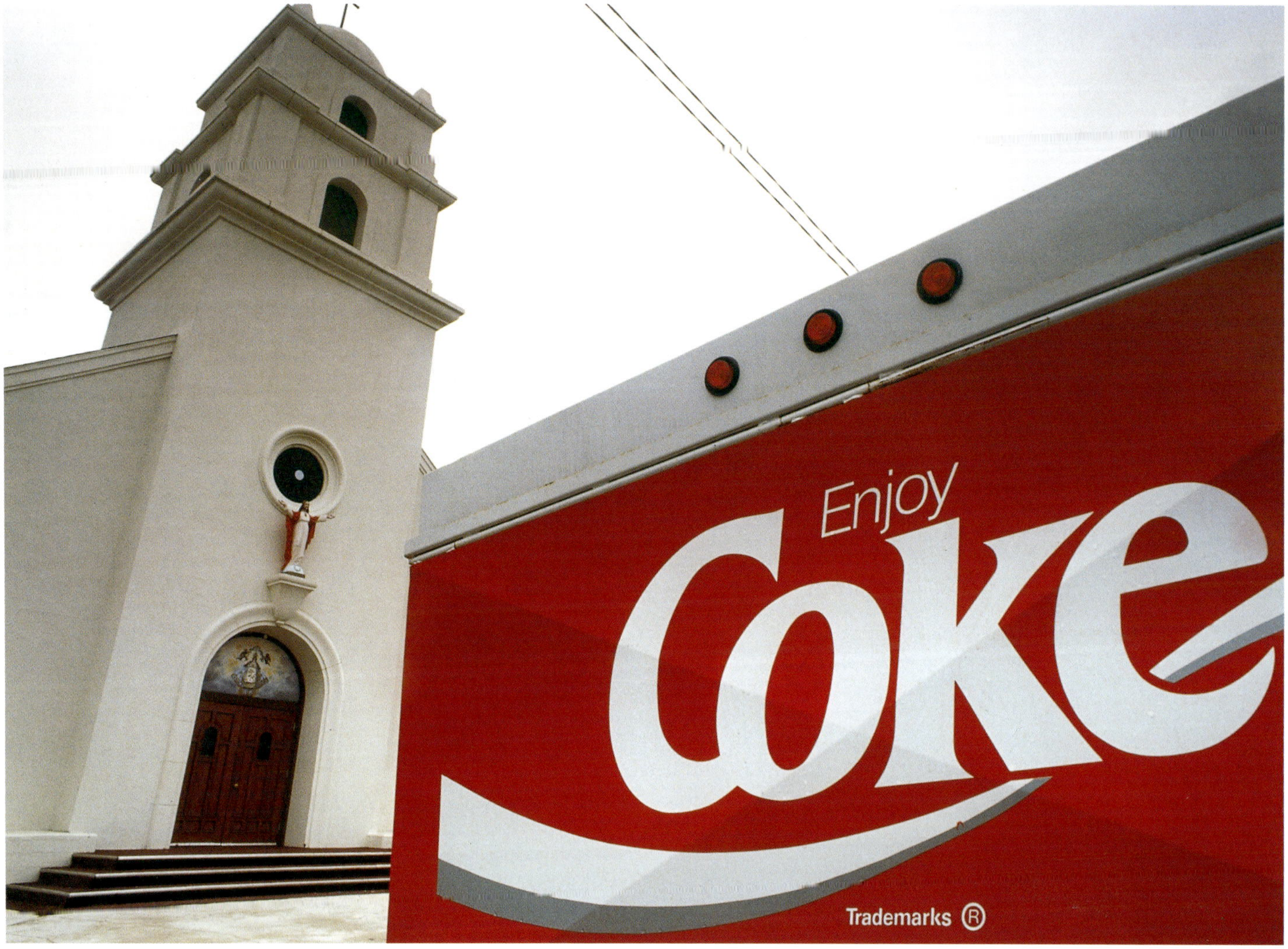

CHURCH AND COKE
Crystal City, Zavala County, Texas

was greeted on my meridian travels — without hostility, paranoia, or insistent curiosity. Never in four years was I greeted with unpleasantness. During that time, I met schoolteachers, wranglers, shopkeepers, farm hands, grain elevator operators, refugees, roughnecks, and people in every imaginable small business. Each was eager to share stories about weather, loneliness, struggle, and adversity, and every place seemed to evoke stories as well.

Inquiring about the painted purple plum on the water tower, I discovered that "Plum Creek" was Lexington, Nebraska's, original name. Later on while exploring the original Plum Creek site, I encountered the story of Nancy Jane Morton who, in 1864, was abducted by a Cheyenne war party after her family was killed in a wagon train ambush. She was tormented by her captors in spite of their respect for her keen rifle marksmanship. With an arrow wound in her thigh, Nancy survived cruelty and starvation during a horseback journey of 1,600 miles in six months through the dead of winter, and eventually was reunited with her people.

At a wonderful Asian restaurant, the "Lucky 7 Cafe" in Winner, South Dakota, I learned of another courageous journey. The cafe's proprietress, Rose, fled the hell of wartime Vietnam by escaping with the boat people, found her way to America's heartland, and started a new life among strangers. Although she didn't dwell on details, Rose's trans-Pacific odyssey must have demanded fortitude equal to Nancy

Morton's over a century ago. Such stories were everywhere.

Faith, Identity, and Self-Expression

For Rose, as well as earlier heartland immigrants, moving halfway across the world to improve one's life was a monumental act of faith. These were men and women at the center of a knowable universe believing that, given time, energy, and God's help, they could master the world. The outrageous demands of the new environment certainly tested that faith.

The dark side to decades of hardship includes a retreat into self and family and a reluctance to accept strangers, new ideas, or alternative lifestyles. Obsessive focus on the past often reinforces a false notion that the old strengths, instrumental in averting past disasters, will save the day tomorrow. Faith in prior solutions inhibits more creative approaches to solving present-day problems along the meridian. The creativity at Bar-S Monogram Industries seemed to be an exception. In Paint Rock, Texas, a rancher's son now ranching himself, shared his concerns about "a hereditary myopia, the inability of fourth generationers to see the land in new ways, to adapt to changing economic realities." In *Dakota: A Spiritual Geography,* Kathleen Norris discusses the origins and persistence of the fortress mentality, as does Osha Gray Davidson in his detailed chronicle of the rise of corporate agribusiness at the expense of the family farm, *Broken Heartland.*

Alongside strategies for adaptation and change, there were expressions of faith and spiritual certitude. In the wreck of an abandoned Texas farmhouse, I found four-inch crosses of woven grass nailed to each bedroom door jamb. They were of a style used for making functional items in tropical Mexico, recently adapted to making dolls and decorations for tourists. The crosses were as intricate, intimate, and artful as the decoration on a traditional Japanese gift box.

I saw the perfect contrast to these delicate grass crosses on a hillside outside of Ballinger, Texas, in the form of a hundred-foot stainless steel cross erected by a local ranch family. Known as the "Studer Cross," this fifty-ton monument dominated the skyline and landscape for a mile or more in every direction. Shiny, tall, and impressively precise, the huge icon glowed brilliant silver in the sun.

These crosses were differing expressions of faith. The Studers' statement in stainless was dramatic, insistent, and public. Besides religiosity, it demonstrated the power of technology and the efficacy of material statement. It was as much about conquest of the environment as about spirituality. Made of natural materials, the grass crosses reflected harmony with the environment. Modestly scaled, privately used, and artfully crafted, their size implied portability. The maker of such crosses was likely a renter or an itinerant, not a property owner. Only landowners build in stainless steel.

Ranchers and farmers historically have mutually contradictory attitudes toward government. At the meridian, a frontier-born distrust of governmental control and regulation is balanced by a dependency on the government to provide for real needs, formerly in protection of settlements through forts and troops, today in the form of agricultural research, soil analysis, weather prediction, and farm subsidies.

Regardless of the ambivalence about government, there's no doubt about patriotism at the meridian. I saw it in modest and in grand statements — American flag mailboxes, the entire roof of a huge North Dakota barn covered by a painted American flag, and in Nebraska, a permanent flag constructed from two-by-fours and firmly anchored on a front lawn. What greater love for the flag than to drape it over the contents of one's barn or to build it as a permanent addition to one's house?

The cemetery on the Standing Rock Indian Reservation and other Native American cemeteries, held ardent displays of the American flag that moved and puzzled me. How could a people, victims of so much unthinkable injustice, heedlessness, and unfairness at our collective hands embrace this nation with such exuberance? Were I the victim, I'm not sure I could have done so. While some of these evocations are attributable to the Plains Indian warrior tradition, I found unabashed enthusiasm for America everywhere, leaving no doubt of the Native Americans' love of country.

■

Defining a unique local identity in a treeless prairie landscape is something of a challenge and a compulsion to communities. Self-identification as the "world capital" of something based on an unusual local product is a sure-fire shortcut to uniqueness in a flat world. Thus, Winner, South Dakota, became the "Pheasant Capital of the World"; Dodge City, Kansas, proclaimed itself "The Cattle Capital of the World"; and Edwards County, Texas, "The Angora Goat Capital of the World." Moderate summer temperatures coupled with year-round artesian water created the "Spinach Capital of the World" at Crystal City, Texas, which inspired a statue of "Popeye," the spinach-guzzling sea captain, in front of City Hall.

North of Lake Oahe, near Pierre, South Dakota, there's a plaque spotlighting the "Centennial Acre 1889–1989" as the central point of the state *and* all of North America. Contrary to these claims, Rugby, North Dakota, defines itself as the "Geographic Center of North America" and has the monument to prove it. But even at Rugby, I discovered that competition stretches geographical truth. Rugby's monument had been moved into town for the sake of "convenience," and the real geographic center was elsewhere. Embarking on a search for the exact center of North America in the inconveniently rural country outside of town, I found it west of Balta, North Dakota, at Latitude 48 degrees 10 minutes north and Longitude 100 degrees 10 minutes west. The center was a white pole, far from shore in the middle of a small, muddy lake in the middle of nowhere.

It isn't surprising that this place of huge skies and unending horizons inspires a search for individual as well as collective identity. In country whose culture has little use for conventional art on canvas or paper, virtually every male can use a welding torch — metal seems to be the medium of choice in the Heartland. I saw many affirmations of folk sculpture welded from scrap auto parts and junk iron. Guarding the street in front of the Anders Body Shop in Stamford, Texas, were a pair of dinosaur skeletons, full-sized works of contemporary "paleontology" skillfully welded from scrap vehicle parts. Not in Kansas, Toto, but near Laverne, Oklahoma, I found a seventeen-foot tall metal sculpture of the Tin Man from the *Wizard of Oz*, built by the owner of the "Log Cabin Cafe and Corner," an ex-oil field worker named Pete Morris.

Unintentional monuments sometimes were the most poetic. In the glaciated terrain of North Dakota, cleared fields are marked with rock cairns that, in the gray fog of daybreak, looked like sacred leavings from a prehistoric race, yet on a sunny morning in May the same cairns nested among carpets of brilliant yellow mustard became a joyful celebration of the labor, persistence, and dreams of the pioneers that built them.

Reflections

During the project, I made seven trips over huge distances, enduring God-awful weather and some of the most dreadful restaurant food in America. I absorbed some increment of wisdom about people, history, and time. Mostly, I came to understand about stories in a deeper way, not in terms of narrative, but in how history flows from the interaction of land and people through story — that whether you touch the land or the people, collective identities emerge to define our myth, shape our thought, and weave the fabric of our very souls.

When starting the project, I held the conventional thought that meridian history was yesterday's stuff of textbooks and yellowed newspapers. But that was only one aspect of the picture. My revelation was in experiencing the enormous mosaic of personal stories that render history an enormous discipline, as large, varied, and deep as its human dimension. And lastly, I learned that history is *now*. Returning to Tuttle, North Dakota, after a year's absence to rephotograph a certain street, the entire block I had photographed seemed to have disappeared. Its buildings had burned down and had been rebuilt with a completely different look. Impermanence is history's handmaiden. In Lexington, Nebraska, I witnessed the demolition of DeBruce's great grain elevators, huge structures that had dominated the townscape for generations. When the wrecking ball stopped, Lexington would never look or feel the same. Witnessing such small events and recording thin slices of geography captures history itself, and the whole notion of photographing the 100th meridian expanded its meaning.

If history is today, what about tomorrow? The future of the meridian is ambiguous, defined by default as much as by action, by what will and won't be. The "will be" is subject to a thousand interpretations, depending on to whom the question is addressed. But in the near term at least, corporate enterprise will continue to absorb family farms, the size and number of small towns will continue to shrink, and services will consolidate at ever-more-distant locations.

In "The Buffalo Commons as Regional Metaphor and Geographic Method," Frank J. and Deborah E. Popper suggest that the very ambiguity of the heartland future can be turned into a plus.[8] Acceptance rather than denial of geographic and climatic realities has proven, in the long term, to be what works most successfully, and that the keys to a productive future for the meridian may be offered by history.

The prairie can be seen as a reservoir of values, a source of national rejuvenation, a refuge from noise, crowds, and intensity, and a place in which to reacquaint ourselves with Nature, with the sky and the earth. One can, in the spirit of the old 1940s tune "Don't Fence Me In," still "ride to the ridge where the west commences, gaze at the moon until we lose our senses."

The landscape and its miracles persist while the meridian people pursue their lives and legislators debate depopulation, revitalization, and resource depletion. Whatever the future of our coastal regions, we must ensure that the meridian endures forever, that it remains a place of sky, wind, earth, a place to get in touch with what William Kittredge calls simply, "the infinities."

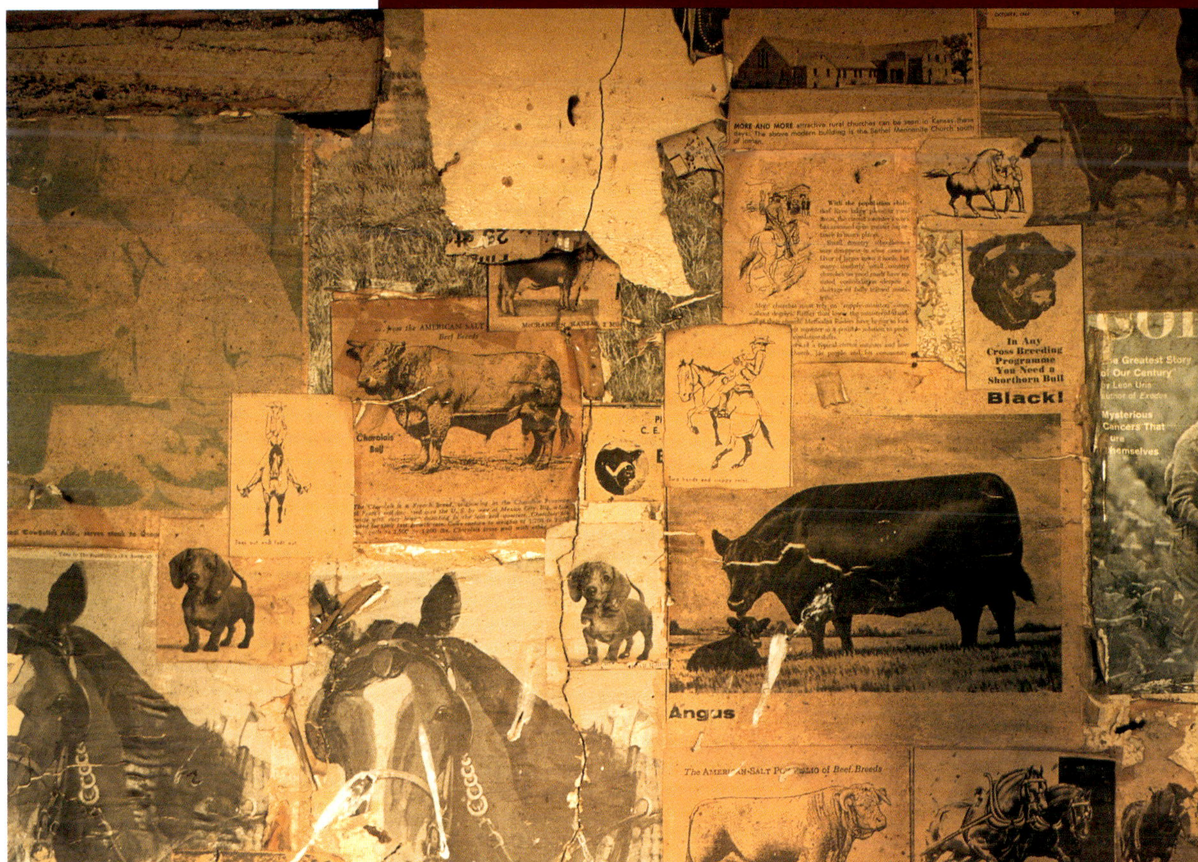

WALLPAPER DREAMS

South of Ness City, Ness County, Kansas

Epilogue

It's been almost a decade since I made my last continental traverse to the 100th meridian, and turned against the grain of our east-west highway system, into the interstices of space and time that we call the Heartland.

What first emerged for me from the elongated blur of highway miles were the generalities: sky, land, beauty, enormity, a sense of history. But the very measure of these epic entities was threatening. I felt like an interloper in a drama being played out on a stupendous stage. Subsequently, those larger abstractions resolved into remembered moments of prairie poetry. The perfect flatness of the land was embroidered with the stitchery of time, geologic and recent. Traceries of fence lines seemed to baste together the land and sky.

Out there under the great sky, it was easy to hear the groaning of the wagon shafts and wheels bringing my stepchildren's great grandparents from Missouri, mile after mile, through prairie mud and dust. Everywhere there was a sense of time, of memories of just yesterday and long ago, all woven into a fabric of myth blanketing this landscape of immense and surreal scale, space, and silence.

In a perversion of scale, these memories recalled Salvador Dali's tiny landscape painting, "The Persistence of Memory," a mini-panorama punctuated by the occasion of limp watches draping the sun-bleached branches of dead trees. Despite its astronomical expansion of Dali's scale, the meridian landscape is draped not with watches, but with the effects of time; with the ghosts of dreams, hopes, and living yesterdays.

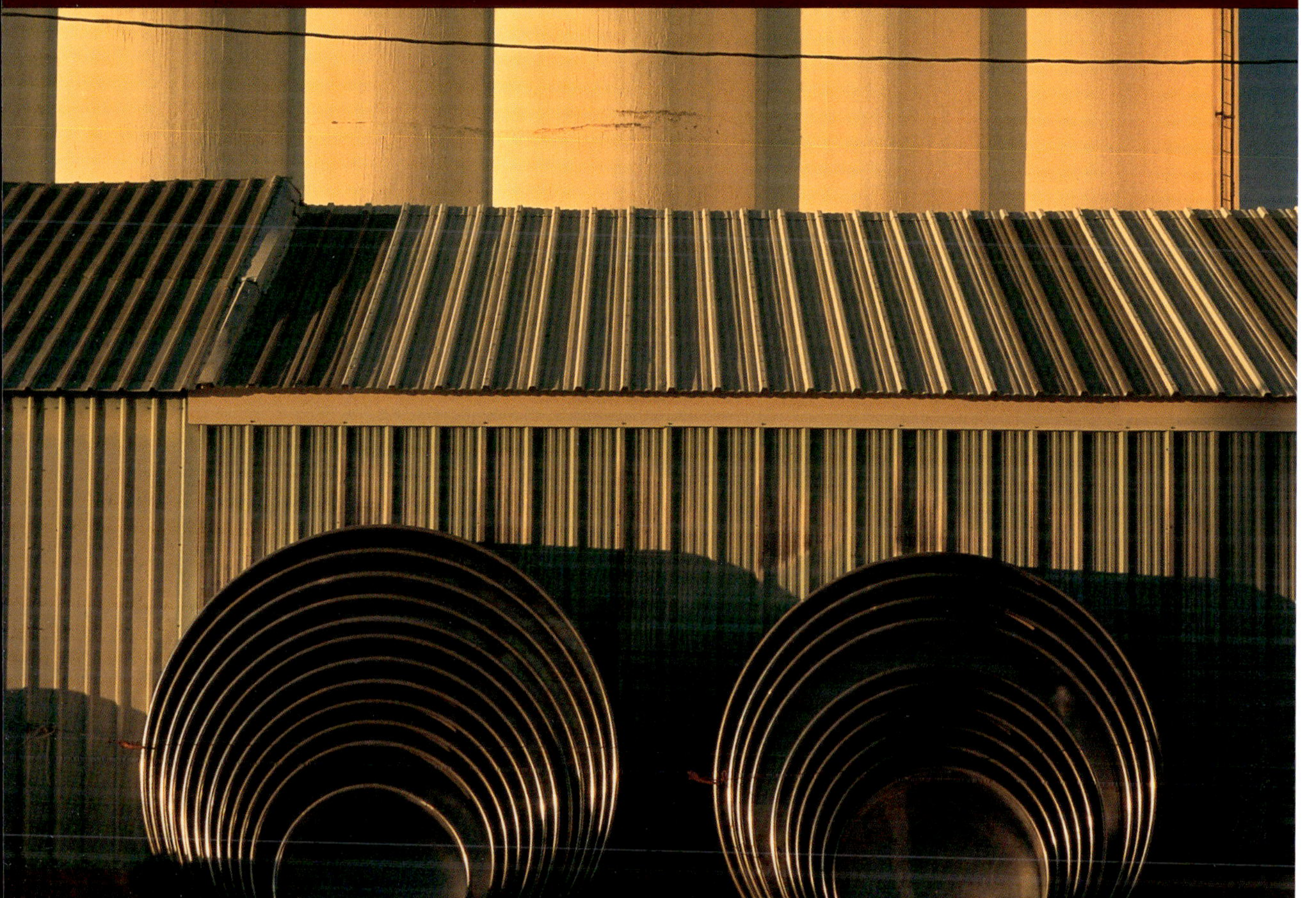

STOCK TANKS AND ELEVATOR
Ransom, Ness County, Kansas

My final revelation was the paradox of the heartland: It's a place where magnitude is a metaphor for itself. Once one moves past the issues of insignificance and solitude, and "gets" what that is about, the once-threatening enormities become a locus where one feels most human, where the presence of elemental forces also yields a gently intoxicating, deeply personal sense of aliveness.

Driving the great length of the meridian's vastness, trying to snatch images from my experience that might somehow make sense to others, I was continually accompanied by the strains of country music. The twangy ballads stayed with me, emblematic of specific moments of memory, and reminding me that I too, after months of wandering, now belonged, in a small but meaningful way, to this larger scene.

It was the refrain of a maudlin, half-drunk ranch hand, bemoaning the inevitability of departure from the worldly feast of sky and prairie, with a request that on his death he be propped up against the jukebox, anchored by sand-filled boots, a stiff drink in his hand. I could get that. One supplies the music as one can. A large sadness came over me when it was time to leave the party.

A Note on the Photography

All photographs for this volume were made by the author with 35mm Nikon F3 camera bodies and Nikkor lenses with focal lengths varying from 20 to 300mm. During the course of the project, autofocus technology was not commonly available, and manual focus lenses were used. The 20mm lens was used great deal, as it seemed to "sweep in" the open space of the prairie, yet permit working with particular subjects fairly close in.

"Fujichrome 100" transparency film was used until the release of Fujichrome RDP II 135 "Provia" (ISO100) film, at which time it became the film of choice for the 100th Meridian project.

Conventional filtration techniques, including warming filters, polarizers, and graduated neutral density filters were employed. A tripod was used wherever possible. There was no digital manipulation of the images.

All images save one were made in a swath two to three counties wide adjacent to or intersected by the line of 100 degrees west longitude.

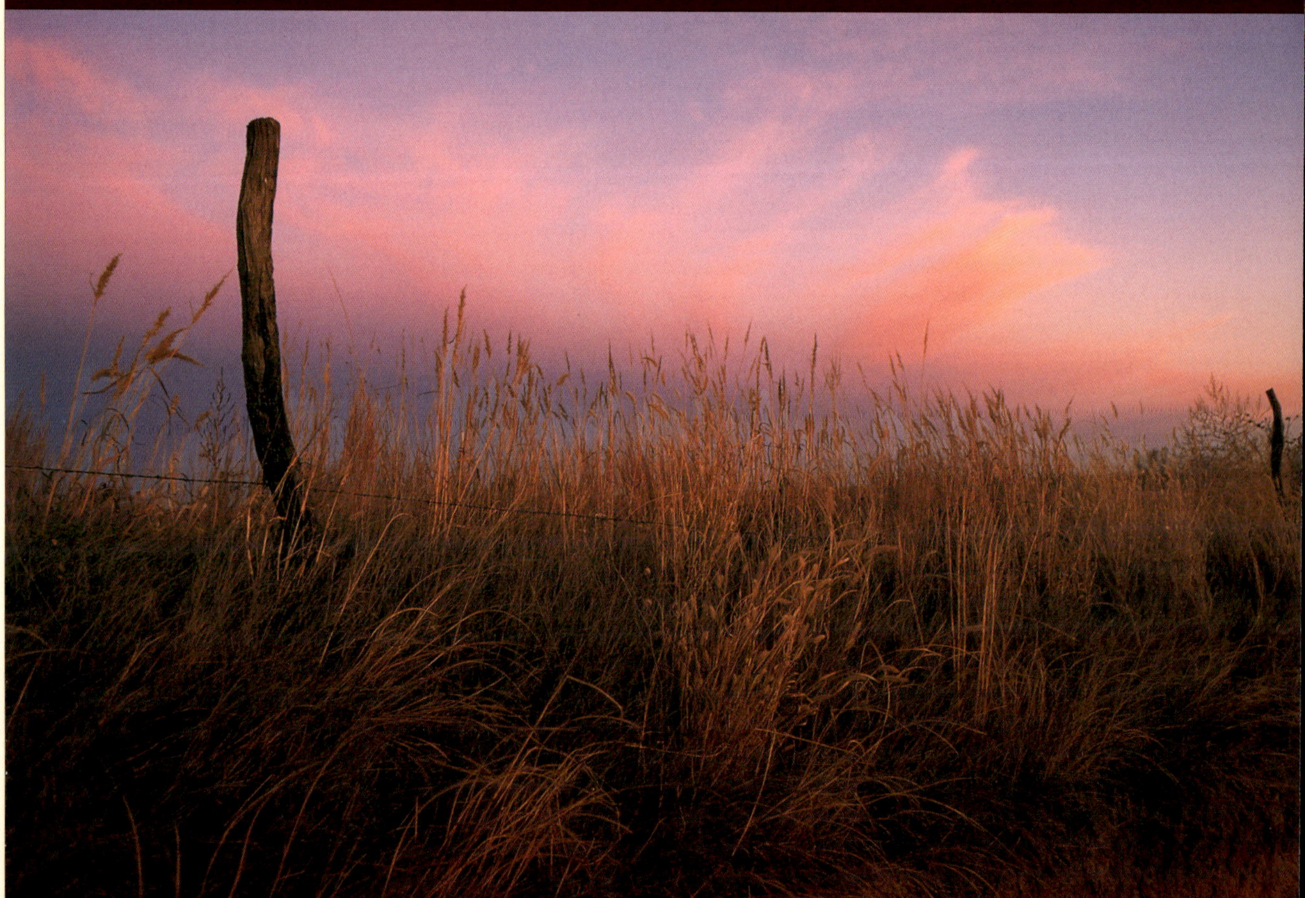

TWILIGHT FENCELINE
Near Nicodemus, Graham County, Kansas

Notes

Epigraph: From Kathleen Norris, *Dakota: A Spiritual Geography* (New York: Ticknor & Fields, A Houghton Mifflin Company, 1993) 157.

1. Wallace Stegner, *Beyond the Hundredth Meridian: John Wesley Powell and the Second Opening of the West* (Boston: Houghton Mifflin Co., 1953), 275.

2. Quoted in William Least Heat-Moon, *PrairyErth (a deep map)* (Boston: Houghton Mifflin Co., 1991), 5.

3. See Janet M. Neugebauer and Walt McDonald, *Whatever the Wind Delivers: Celebrating West Texas and the Near Southwest* (Lubbock: Texas Tech University Press, 1999).

4. Laura Bush, "Foreword," in ibid., xi.

5. See Terry Evans, *The Inhabited Prairie,* with an essay by Donald Worster (Lawrence: University Press of Kansas, 1998).

6. Quoted from *The Inhabitants* by Wright Morris, in ibid., ix. See also Mari Sandoz, *Love Song to the Plains* (New York: Harper and Row, 1961); Linda M. Hasselstrom, "A 'Love Song to the Plains'," an essay monograph published by the Mari Sandoz Heritage Society and edited by John R. Wunder (Lincoln: Center for Great Plains Studies, University of Nebraska, 2004), 6–21; and the journal by the same name published by the Center for Great Plains Studies that features undergraduate and graduate student poetry and short fiction.

7. Quoted in Least Heat-Moon, 4.

8. See Frank J. and Deborah E. Popper, *The Buffalo Commons as Regional Metaphor and Geographic Method* (unpublished in 2000, but accepted for publication by *Geographical Review*).

Index

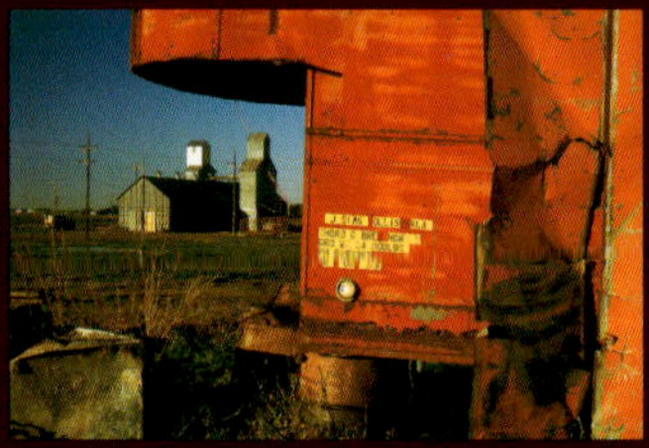

The authors and Texas Tech University Press are deeply grateful to The CH Foundation, without whose generous support this work and the books in the Plains Histories series would not have been possible.

The story of the North American Plains is the story of indigenous peoples and the multitude of other cultures that settled near them. It is the story of each culture's adaptability, whether that culture be ancient farmers on the Platte, cattle ranchers of the High Plains, or entrepreneurs of the twenty-first century.

Plains Histories is devoted to illuminating and defining the Great Plains, its peoples, and its environment through original monographs, edited collections, biographies, memoirs, and comparative studies.

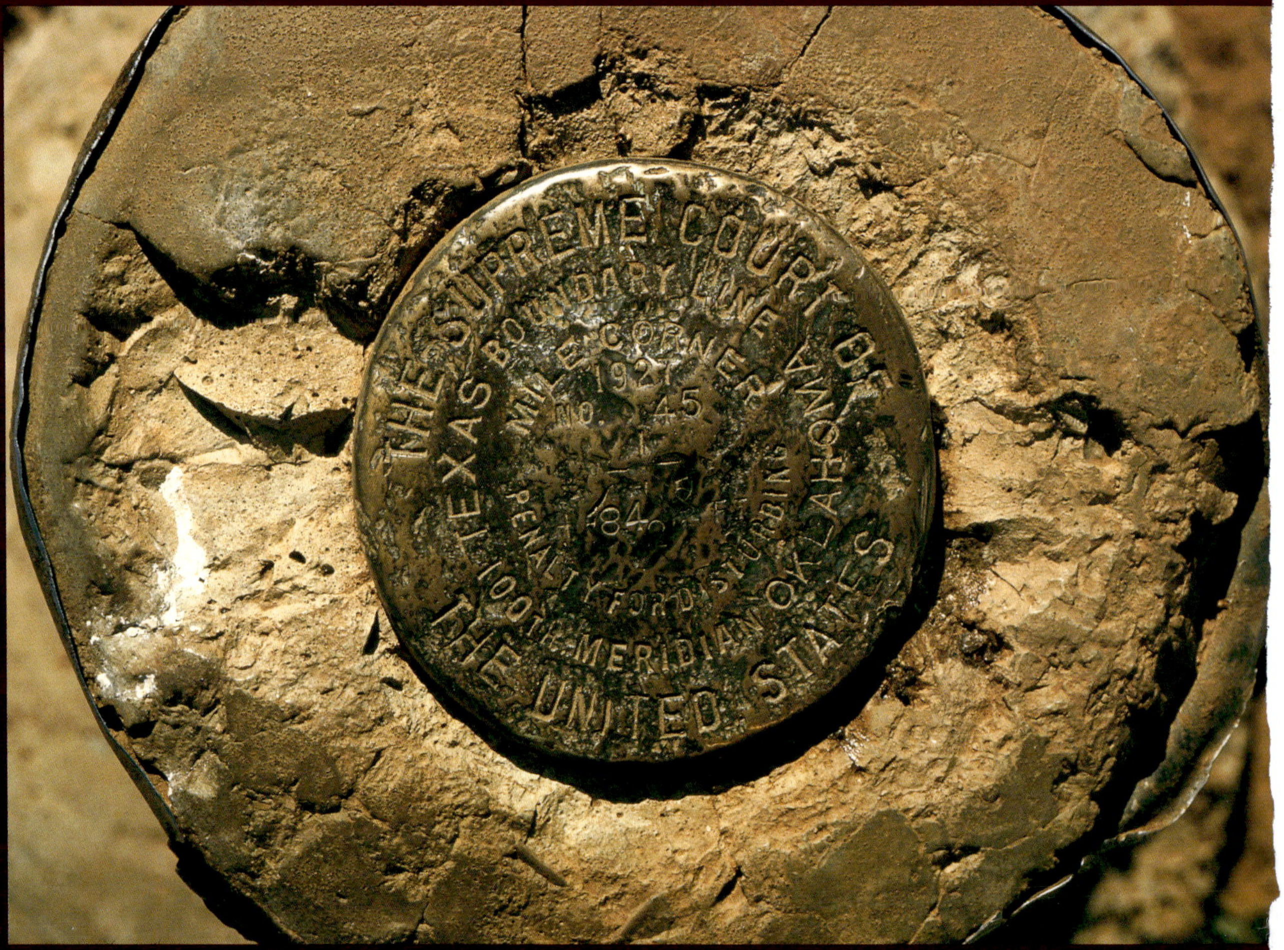

100TH MERIDIAN MARKER
Texola, Beckham County, Oklahoma

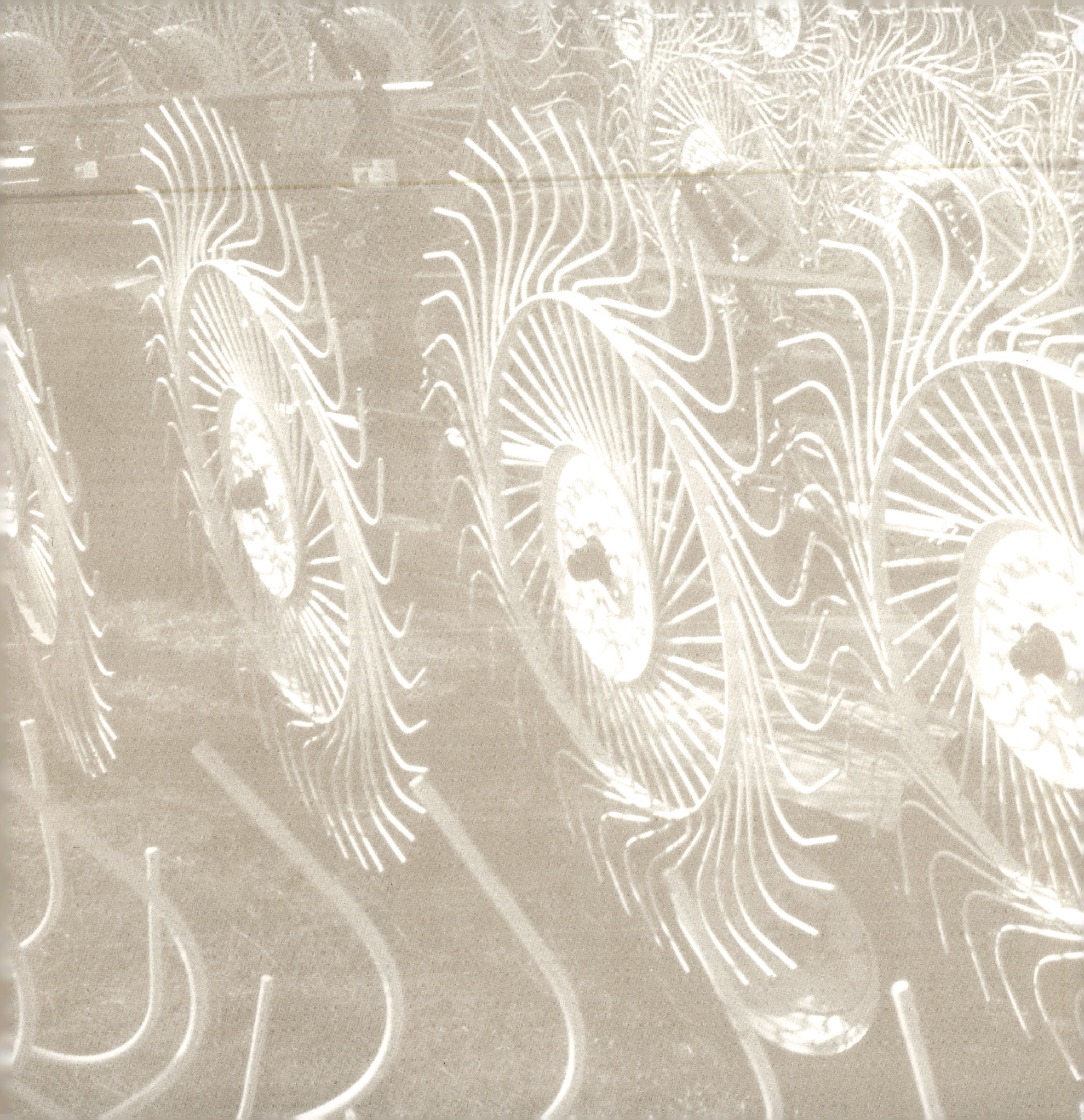